Flowers of
Point Reyes National Seashore

Flowers of
Point Reyes
National Seashore

Roxana S. Ferris

Illustrations by Jeanne R. Janish

University of California Press
Berkeley, Los Angeles, and London

To John Thomas Howell whose classic,
Marin Flora, has aided me immeasurably
in working with the plants of Point Reyes
National Seashore

University of California Press
Berkeley and Los Angeles, California

University of California Press, Ltd.
London, England

ISBN 0-520-01694-7 (paper)

Library of Congress Catalog Card Number: 70-111308
Printed in the United States of America

34567890

Contents

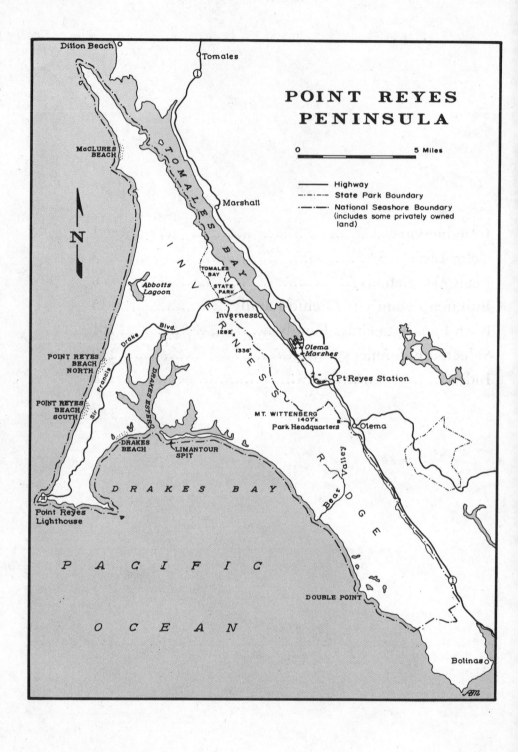

INTRODUCTION

This is your wildflower book for Point Reyes National Seashore. The pictures of the plants selected from among the many that grow on the Seashore are accurately reproduced, usually at somewhat less than natural size. The inset sketches of growth habit will have a scale drawn beside them to indicate size. The text will tell you something about the plant: its flower color; the plant family it belongs to; what kind of place it grows in; its scientific name and other related information. The arrangement of the plant families is the same as that in *A California Flora* by Philip A. Munz.

If you wish to identify flowers by color only, all species included in the book are arranged in color groups (blue, red, yellow, etc.) in the tabulation on the following pages. Page numbers are given after each species listed to help locate the description of the plant and its picture. Some of the plants illustrated in the book are not included in the color lists: grasses, sedges, rushes, plants without petals or with such inconspicuous petals that there is no use trying to see them. To those who wish to know more about the flowers there is a list of books in an appendix about plants growing in the West. Some of these books are technical and some are of a popular nature. Some are for the Pacific States and some are limited to plants that grow in the central part of California.

COLOR LISTS

Red

Anagallis arvensis 31
Aquilegia formosa truncata
  (with yellow) 15
Castilleja franciscana 43
Castilleja wightii 43
Orthocarpus pusillus (maroon).............................. 44
Tellima grandiflora 53
Vicia gigantea (maroon) 68

Blue

Anagallis arvensis azurea 31
Calochortus tolmiei
  (dull bluish white) 89
Camassia quamash linearis 92
Ceanothus gloriosus gloriosus 72
Ceanothus gloriosus porrectus 72
Ceanothus thyrsiflorus 72
Delphinium decorum 15
Gilia capitata chamissonis 38
Lupinus arboreus (purplish blue) 62
Lupinus bicolor (with white) 63
Lupinus variicolor (with white) 63
Lupinus polyphyllus (purplish) 63
Myosotis latifolia 42
Navarretia squarrosa 38
Nemophila menziesii 40
Viola adunca (purplish) 21

Yellow Orange

Abronia latifolia 31
Agoseris apargioides 77
Agoseris grandiflora 77
Amsinckia spectabilis (orange) 42
Brassica kaber 24
Camissonia cheiranthifolia 69
Camissonia ovata 70
Castilleja wightii (pale yellow) 43

Pink, lavender or purple

White or cream

PLANT ASSOCIATIONS

The scenic beauty of the Point Reyes National Seashore is determined by the underlying geological structure; the trees and shrubs, the grasses and herbs, the ferns, lichens, and mosses clothe this form. Point Reyes Peninsula has a land mass reaching in places to an altitude of 1,400 feet and is bounded on one side by the Pacific Ocean and on the other by Tomales Bay. Within the 53,000 acres of the peninsula there is a great variety of habitats, the kinds of places where the plants grow. Sand dunes and saltmarshes, freshwater lagoons and sag ponds, forests, and grasslands have their own characteristic plants as well as several other plants that are equally at home in more than one habitat. Much of the land of the peninsula is privately owned and not open to the public but much is now state or federal property. Eventually even more will be National Seashore where people may wander at will and see for themselves where plants grow and what they look like.

During a nature walk at McClures Beach on one of the last days of June, 79 kinds of plants were counted growing beside the trail from the top of the bluff to the beach sand. Some of these plants were not in bloom and about one-third of them were just pastureland weeds, but the gorgeous display of the lavender, crimson-spotted cups of *Clarkia* (formerly known as godetia and sometimes called "summer's darling") alone made the walk worthwhile even if no other wildflowers had been seen. On the year of our walk, the yellow bush lupine was mostly gone but pale yellow paintbrushes and occasional plants of the cobweb thistle, which has all its parts draped with cobwebby hairs, demanded attention.

Associations of plants that can grow together under the same conditions are called plant communities. The plant communities found in Point Reyes National Seashore are listed below under headings that are descriptive of the local conditions of this area. Rarely are the lines between one and another as completely sharp as that shown between the plants of the "saltmarsh" and those of "grassland" or "farmside roads and pastures" that are immediately beside the marsh. Even within a plant community so clearly defined as "bishop pine forest" there are "islands" for instance of treeless swales of Nootka reedgrass and common rush which are just another type of "grassland," or it may be that a "mixed evergreen forest" with madrone, tanbark oak, California bay, California coffeeberry, and chinquapin encroaches on the north slopes of the pure stands of the pine. With all these exceptions—and nature is full of them—it still seems worthwhile to list the most prominent plant communities.

Your preliminary botanizing on Point Reyes Peninsula can be done by automobile, for the existing roads either touch, or are in plain sight of, every kind of plant community, be it saltmarsh, freshwater lagoon, Douglas-fir forest or mixed evergreen forest and coastal scrub. Much additional information on the subject can be found in the introduction to *A California Flora* by Philip A. Munz and that of the *Marin Flora* by John Thomas Howell.

Just a very few characteristic kinds of plants are given in the following local list of habitats, and those that are given, for the most part do not mingle with the vegetation of the adjoining habitats. In passing I might point out that poison oak is commonly found in all kinds of places except "saltmarsh" and possibly "beaches and dunes." Take a botany book and notebook along on your walk and search for less common plants that are restricted to their own plant communities, instead of appearing in many like the omnipresent coyote bush.

Beaches and dunes. Beach morning-glory, yellow sand-verbena, sea-rocket, beach sweet pea.

Saltmarsh. Pickleweed, Pacific cordgrass, alkali heath (*Frankenia,* not illustrated), saltgrass (*Distichlis,* not illustrated), marsh-rosemary.

Ledum swamps and freshwater marshes. Deerfern (*Blechnum spicant,* not illustrated), ledum, salmonberry, twinberry, silverweed, marsh beaked-buttercup, bog lupine, serviceberry (*Amelanchier,* not illustrated), round-leaved psoralea, California waxmyrtle. The last-named species is also common in grassy openings in the bishop pine forest.

Freshwater lagoons and inland ponds. (Surrounding shrubs much the same as the preceding) cat-tail (*Typha,* not illustrated), common tule, marsh pennywort, water parsley, smartweed, silverweed, willow herb (*Epilobium,* not illustrated).

Coastal bluffs. Seaside daisy, gumplant, California phacelia, bearberry, live-forever, lizard-tail or seaside woolly-daisy, coast buckwheat, coast rockcress. ("Coastal bluffs" are often included in some books with "beaches and dunes" under the name "coastal strand." Many plants stray from one area to the other.)

Coastal scrub. Thimbleberry (often in other habitats), California sagebrush (*Artemisia californica*, not illustrated), osoberry, giant vetch, cow-parsnip (often in other habitats), flowering currant.

Grassland. (Headlands and the slopes away from the ocean) tidy tips, cream cup, brodiaea, common rush, yellow mats (*Sanicula arctopoides*, not illustrated), morning-glory, sun cups, Nootka reed-grass, coast lotus, cardionema, checker-bloom, shooting stars, narrow-leaved mule's ears, California mahonia.

Streambanks. Red alder, willow, California spikenard, big-leaved maple, five-fingered fern (none of the preceding illustrated), gooseberry, hazelnut, brookfoam.

Farmside roads and pasturelands. Poison-hemlock, milk thistle, wild radish, Scotch broom, burclover, bird's-foot trefoil, cheeseweed.

Mixed evergreen forest with shrubs. This is rather an indefinite zone on the lower north- and east-facing slopes of both "bishop pine forest" and "Douglas-fir forest." California bay, buckeye, tanbark oak (none of these illustrated) characterize it along with such shrubs as cream bush, red elderberry and blue elderberry, blue blossom, California coffeeberry (*Rhamnus californica*, not illustrated), hazelnut, sticky monkey-flower, California fescue, which are also found within the borders of the forest.

Bishop pine forest. Vanilla grass, bedstraw (*Galium*, not illustrated), salal, coffeeberry (*Rhamnus*, not illustrated), blackberry, star flower (*Trientalis*, not illustrated), redwood rose (*Rosa gymnocarpa*, not illustrated), glory-mat, yerba buena.

Douglas-fir forest. Red elderberry, blue blossom, coffeeberry, blackberry, thimbleberry, redwood rose (not illustrated), forget-me-not, ladyfern (*Athyrium filix-femina*, not illustrated), leather fern (*Polypodium scouleri*, not illustrated), twinberry.

BUTTERCUP FAMILY
TO
ORCHID FAMILY

COLUMBINE (*Aquilegia formosa truncata*)
Buttercup Family

Columbines are favorite wildflowers, subjects of many a line of poetry. Several kinds displaying many different colors are found over the northern hemisphere. Our columbines grow on bushy slopes and streamsides, especially toward the southern end of Point Reyes National Seashore. Though fairly common they never grow in great masses and it is with a real feeling of discovery that one finds the red and yellow flowers dangling on their slender stems in some shady place. The flowers are really worth examining closely. The red sepals spread at right angles to the stem. The petals have short yellow lobes erect above the circle of sepals, and the remainder of the petal is drawn back into red tubular spurs with nectar in the tip which lures and feeds the long-tongued hummingbirds and butterflies. The stamens and stigmas extend beyond the upper lobes of the petals in a tight cluster ready to brush against the nectar seekers as they feed.

COLUMBINE

COAST BLUE LARKSPUR (*Delphinium decorum*)
Buttercup Family

As its name implies, this larkspur grows only along the coast and can be found from Santa Cruz to Mendocino County. Its deep blue single-spurred flowers are as few as two and not more than six, and the hairy stems are usually less than ten inches in height. Larkspurs (there are many kinds growing in the West) are poisonous to cattle. The coast larkspur does not grow in great masses as the lupines and poppies often

COAST BLUE LARKSPUR

MARSH BEAKED-BUTTERCUP

SHINY-LEAF MAHONIA

do, so do not expect to find them in abundance. They will be scattered in the coastal scrub or on the open grassy slopes.

MARSH BEAKED-BUTTERCUP
(*Ranunculus orthorhynchus platyphyllus*)
Buttercup Family

In the springtime on the Point Reyes Peninsula, the California buttercup (not illustrated) can be seen in abundance in every open grassy field. Much less common but more rewarding to find is the beaked-buttercup. This plant has flowers of buttercup yellow about the size of a silver dollar. It grows in the soggy soil in marshy depressions near the ocean, and the stems loop about the mass of sedges, grass, and shrubbery to a considerable length. If you have the opportunity to compare the seed heads of both buttercups, you will find that the seeds (achenes) of the buttercup of the marshes have conspicuous straight points instead of very short, curved ones.

SHINY-LEAF or CALIFORNIA MAHONIA or CALIFORNIA BARBERRY
(*Mahonia pinnata*)
Barberry Family

Shiny-leaf mahonia grows in the coastal counties from Baja California to Oregon but never in great stands. It is a shrub of open spaces and grows on rocky outcrops, steep grassy hills, or more rarely in an open meadow or in the forest. The clumps of stems are often not more than a foot or so high and definitely mounded. The deer like it for browse and keep it well pruned along with some of their other favorite shrubs. The flowers are buttercup yellow. The berries are blue with a bloom over the surface and are sometimes used to make a very tasty jelly.

MALLOW or CHEESES (*Malva nicaeensis*)
Mallow Family

This plant is a pasture and farmyard annual weed that may persist and display its fluted kidney-shaped leaves all year round. The washed-out lavender flowers on short stalks at the bases of the long-stalked leaves are miniature replicas of the wild hollyhock or checker-bloom that still is so abundant in spring in open countryside that has not been overgrazed. Seeing the circular flattened fruit, each marked off in segments, may give a nostalgic twinge to those of us who as small children served up and ate these "cheeses" when we were playing house.

WILD HOLLYHOCK or CHECKER-BLOOM (*Sidalcea malvaeflora*)
POINT REYES SIDALCEA (*Sidalcea rhizomata*)
Mallow Family

Two species of checker-bloom or sidalcea are pictured here. If we include all the named variations, one (*Sidalcea malvaeflora*), is the commonest species in California; the other, the Point Reyes sidalcea, is one of the rarest, being found only in marshy places on the Point Reyes Peninsula and as far north as Point Arena. The Point Reyes sidalcea has a weak stem supported by the adjoining sedges and bunch-grasses. It has a long creeping rootstock, and its stems and leaves are almost smooth, but the calyx of each of

MALLOW

WILD HOLLYHOCK

POINT REYES SIDALCEA

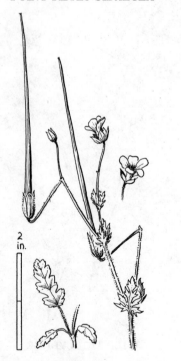

CLOCKS

the pretty pink flowers is quite hairy.

The checker-bloom, or wild holly-hock, may be seen at Point Reyes Beach with a half dozen or so other kinds of wildflowers, completely flattened by the winds of the Pacific into a turf on the fixed, or stationary, dunes. It is equally at home in more protected places and the graceful wands, almost a foot high with pink blossoms that look exactly like small hollyhocks, are easily spotted even from a moving car. The smaller flowers that are occasionally found lack pollen in the rudimentary anthers.

CLOCKS or FILAREE *(Erodium botrys)*
FILAREE *(Erodium cicutarium)*
CRANESBILL *(Geranium molle)*
Geranium Family

All three are weeds from Europe. Both filarees pictured here paint the short grassy hillslopes with pink, and in addition to this attractive floral display are also good forage. The pink-flowered cranesbill, a much less common flower, is more apt to be scattered in the lush grass along farmside roads and adjacent meadows and gardens where one may find also still another weedy species of filaree (*E. moscha-tum*, not illustrated) having washed-out pink flowers. The icepick- or stiletto-shaped fruits, whether the ones on the short fernlike leaved *E. cicu-tarium* or the ones three times as long that grow on *E. botrys*, have a most in-teresting way of spreading and plant-ing seeds. There are five seeds on each "icepick," and as they become dry they spring away from the long central col-umns to which they are attached. As soon as released the long tail on the seed rapidly winds in a spiral, pushing the sharp-pointed seeds into the ground. The spiral will unwind if wet with fog or rain but will dig the seed in further when the sun comes out to dry it. The seeds of the little cranesbills operate on the same principle but not in the dramatic fashion of the filarees—just a turn or two instead of several twists all the way to the head of the seed. Another cranesbill with more divided leaves *(Geranium dissectum)* is often found.

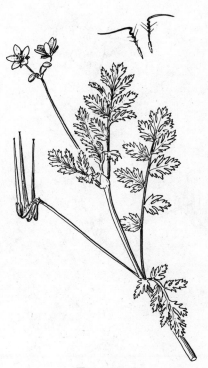

FILAREE

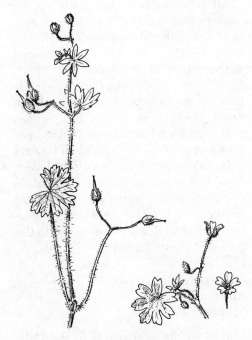

CRANESBILL

YELLOW OXALIS *(Oxalis pilosa)*
Oxalis Family

Open grassland and coastal scrub are the places where this weak-stemmed small yellow-flowered plant is to be found. The first time one notices it in walks around the hills, there is an impulse to weed it out, because it so much resembles the garden weed that is so hard to eradicate. They are different but not obviously so. The stems do not root at the joints as do those of the garden pest, and the flower petals are longer.

MEADOWFOAM *(Limnanthes douglasii sulphurea)*
Meadowfoam Family

All meadowfoams are early spring annuals that need a rainy winter to present a good floral display. They thrive in depressions in fields and pastures where rainwater collects and stands for a week or so, thus giving the seeds time to germinate. Later, usually in April, these depressions are painted with creamy white flowers on the low-growing plants. The individual flowers have white petals with yellow bases. Near Point Reyes itself a completely yellow-flowered form (var. *sulphurea*) predominates. The typical form is found in the Olema marshes and in the area of Abbotts Lagoon.

CALIFORNIA MILKWORT *(Polygala californica)*
Milkwort Family

Most of the year plants of the California milkwort are apt to be completely overlooked. They are perennials. The stems spread from the base so it would be said that the plants are

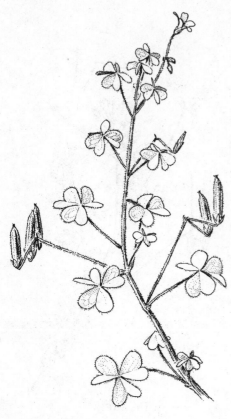

YELLOW OXALIS

MEADOWFOAM

a few inches wide rather than tall. The oval leaves, about one-third of an inch long, are dark green. When the plants bloom in early summer the terminal cluster of two to four bright rose-purple flowers are very eye-catching against the dark green leaves. The first impression is that they are of the pea family because of their somewhat pea-flower shape. Besides these showy flowers there are small dull blossoms that never open at the base of the plant. They are like the nondescript blossoms at the base of some violet plants and just as important. In both plant genera these obscure flowers are self-fertilized and set good seed. There are many kinds of milkworts in the eastern and southern states but very few on the Pacific slopes. The sap is not milky as the name seems to imply. This name was given, so it is said, because some species supposedly could increase the flow of milk in the animals that ate them.

CALIFORNIA MILKWORT

BLUE VIOLET *(Viola adunca)*
REDWOOD or EVERGREEN VIOLET
(Viola sempervirens)
Violet Family

The specific name of the redwood violet means "ever living," and the firm-textured leaves can be found, with some searching, growing in loose mats in the forest duff in the bishop pine or Douglas-fir forests. Though the Point Reyes National Seashore is not a place where redwoods grow, several of the characteristic plants of the redwood forests — such as these violets, the redwood rose, vanilla grass, trillium *(T. ovatum)*, and wild ginger *(Asarum caudatum)* — have moved into other kinds of forests. These yellow violets appear in spring and are usually gone

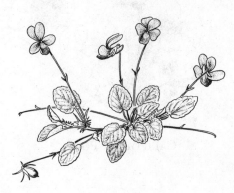

BLUE VIOLET

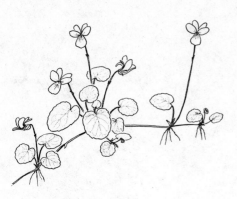

REDWOOD VIOLET

by June, leaving only the rounded leaves pressed against the brown background of needles. The blue violet is not so selective in the places where it grows as is the redwood violet. Throughout its widespread distribution from Alaska to the eastern seaboard there is considerable variation in the shape of the spur, in the shades of purplish-blue in the flower, and also in its size. Plants that grow on the ocean bluffs or along roadsides may look as neat and compact as the plants of the sweet violet of gardens. Those in the pine forest are long stemmed and spread loosely in the tall grass.

CALIFORNIA POPPY

CALIFORNIA POPPY (*Eschscholzia californica*)
Poppy Family

What can be said about California's state flower? It surely is the flower that everybody knows. Very many forms of the California poppy have been thought to be sufficiently different to deserve scientific names, but the one on Point Reyes Peninsula is like the original one that was collected on the dunes of San Francisco by the naturalists of the Kotzebue expedition in 1816. The botanist, Chamisso, gave the name *Eschscholzia* to the poppy to honor his fellow naturalist. The plants are low-growing as is to be expected in this windswept countryside; the leaves are a bluish green and the beautiful satiny orange petals are usually shaded to a paler tone at the outer margins.

CREAM CUP *(Platystemon californicus)*
Poppy Family

Dull green foliage of straplike leaves, stems and flower buds with spreading white hair that is long enough to be easily seen, and pale cream petals shaded to a light yellow at the central mass of stamens give you the picture of cream cup. It is an annual of sunny places and can be seen in the old pastures around Point Reyes National Seashore Headquarters. Grassy headlands from Limantour Spit to Double Point have them sprinkled around in the grass but often dwarfed with but a single flower to a plant. The fruiting part resembles the frayed end of a piece of twine. The seeds are set lengthwise in the dry lightly twisted "strands" which are the several carpels of the fruiting body.

CREAM CUP

ROCK-CRESS *(Arabis blepharophylla)*
Mustard Family

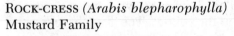

The rose-purple flowers of this perennial of the thin soil of bluffs and headlands can be seen as early as February. The plants are more often a half a foot than a foot high. Most of the hairy green leaves are basal but a few are scattered along the stems up to the flowering part, or inflorescence. It would make a fine rock garden subject. A few nursery catalogues have listed it from time to time. Geographically it is very limited and so far has been found only from the northern arm of Monterey Bay to Sonoma County.

ROCK-CRESS

WILD MUSTARD

WILD RADISH

WILD MUSTARD (*Brassica kaber*)
WILD RADISH (*Raphanus sativus*)
Mustard Family

Wild radish contributes much toward making the farmside roads, pasturelands and vacant lots beautiful in the spring. Like the yellow mustard with which it grows, it is an annual. The plants may grow as high as three feet and are much branched. The branches are loaded with blossoms that vary in color from white, pale yellow and buff to rose, each petal with dark purple veins. The rounded pods do not split open and are slightly constricted between the seeds. I have often thought that the young pods could be made into excellent party pickles. After all, this plant is the ancestor of the cultivated radish and certainly not poisonous. The flowers of wild mustard are all bright yellow. The flattish pods open with age. More than one species of the mustard grows on Point Reyes Peninsula.

SEA-ROCKET *(Cakile maritima)*
Mustard Family

The sea-rockets grow on the sandy beaches right at the edge of high tide. One *(C. edentula californica)* is a native but so inconspicuous it might not even be noticed. *C. maritima*, a recent introduction, has attractive pink flowers and succulent lobed leaves which are so abundant that the large annuals dot the sand with bluish-green masses of foliage. I do not know when it first appeared on the beaches of the central part of California, but E. L. Greene in his *Flora Franciscana* (1895) writes of his surprise that the European *Cakile maritima* was not present with other plants from that source. Fortunately this "weed" is limited to loose sand and it will not become the curse that introductions such as Australian fireweed (Erechites), milk thistle *(Silybum marianum)*, and star thistles (some *Centaurea* species) have proven to be.

SEA-ROCKET

WALLFLOWER *(Erysimum concinnum)*
Mustard Family

From February at least through April the cream-colored wallflower is the prizewinner in the flower garden of the coastal strand. It is not as brightly colored as the purple rock-cress or the big-flowered Lasthenia, but it would win on such points as best display of blossoms, compact form, or good placing in the flower bed. The blunt straight pods angle upward or radiate about the upper stem like spokes of a wheel. It grows from Point Reyes northward to the Oregon coast.

WALLFLOWER

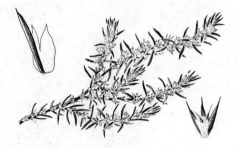

CARDIONEMA

FIELD CHICKWEED

CARDIONEMA (*Cardionema ramosissima*)
Pink Family

Looking straight down on plants of cardionema one almost expects this pinwheel of a plant to spin, so close to moving do the radiating stems of gray-green and white seem. The narrow, short leaves are closely set on the stem, and attached to each side of the leaf is a shining, thin papery wing. There is not much difference in the appearance of the plant when it is in flower or when merely in leaf. The blossoms are purely functional, and it takes a hand lens to distinguish the parts of the flower. Cardionema is found in thin soil on hilltops near the sea or bluffs above the beach. This is another example of a species that is at home on the coast of Chile and also grows as a native on the Pacific Coast of North America from Mexico to Washington.

FIELD CHICKWEED (*Cerastium arvense*)
Pink Family

This is related to the snow-in-summer (*Cerastium tomentosum*) that is much planted in flower gardens. The flowers with their notched white petals are just as large but the leaves are not so silvery white. In contrast to the other chickweed illustrated (*Stellaria littoralis*) which is so limited in its range, this one grows in many places. On the west coast it is found, in some of its variable forms, in high mountains as well as the seacoast from here to Alaska. It grows also on the Atlantic side of the United States, Canada, northern Europe, and Asia. Look for it on bluffs above the beaches or in the grassy headlands.

Coastal Catchfly or Campion
(Silene scouleri grandis)
Pink Family

This plant does not have the large brilliant red flowers of *Silene californica* which hopefully will be found within the boundaries of Point Reyes National Seashore, but the tall, almost velvety stems, which may be as high as two and one half feet, catch the eye. The cleft petals, which vary from greenish white to rose, stand at right angles to the sticky calyx. This plant is of infrequent occurrence. It grows along the ocean bluffs from San Mateo County to British Columbia.

Shore Chickweed or Starwort
(Stellaria littoralis)
Pink Family

When John Torrey studied the first collection ever made of this plant and found it to be a new kind, he gave it the name *Stellaria littoralis*. The plant specimen had been collected by John Bigelow, botanist of the Pacific Railroad survey, who in April, 1854, took a week off to botanize around Point Reyes. The shore chickweed and several other plants are among his "firsts." The word "littoralis" means "of the seashore," and this relative of the weedy chickweeds of your garden lives up to its name. Its weak stems clamber through the rich green tangle of herbs and low shrubs by freshwater lagoons or in gullies in the bluffs directly above the beaches, which are the only places it is found. The area where it grows is limited also. It has been found from the bluffs of San Francisco along the coastline as far north as Humboldt County.

Coastal Catchfly

Shore Chickweed

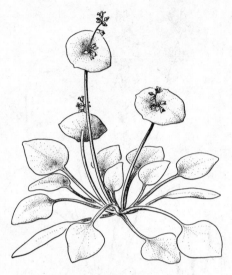

MINER'S LETTUCE *(Montia perfoliata)*
Purslane Family

A walk in the woods in early spring in the filtered light of a trail will almost always lead to one of nature's most effective annual ground covers. Plants of miner's lettuce, with succulent round leaves that completely encircle the stems, massed together can conceal the ground beneath. The rather ridiculous little bouquet of small white flowers atop this leaf is just another attraction in favor of this little plant. It is edible and probably both miners and Indians made use of it for pot greens, but don't expect it to sustain life if you are lost in the woods.

MINER'S LETTUCE

SIBERIAN MONTIA *(Montia siberica)*
Purslane Family

The Siberian montia, along with other moisture-loving plants like tinker's penny *(Hypericum anagalloides)*, yellow-eyed grass, monkey-flowers, and others, will be found in wet ditches, ledum swamps, and moist edges of lagoons and streams. Its leaves and stems are of that lush, succulent texture of miner's lettuce, but the stem leaves are not fused together and the basal leaves are pointed and taper to a long slender stalk. The flowers are pinkish or white either with deeper pink pencilling. As you can guess from the species name it is another plant of the north that has extended its range down the cool fog-drenched coast.

SIBERIAN MONTIA

COAST BUCKWHEAT (*Eriogonum latifolium*)
Buckwheat Family

Like most of the other perennial buckwheats which seem to thrive in adverse conditions (thin soil, rocks and cliffs, and full sun), the coast buck-wheat clings in full force of the ocean wind by its woody roots to the sea bluff and rocks. This plant with full heads of flowers, white or tinged with rose, and the basal clusters of wavy-margined leaves densely felted with white hairs (at least on the underside), takes its place in the natural rock gardens of the Point Reyes National Seashore. Blossoming as it does in summer and later after the flowers have dropped from the coast wallflowers, rock-cress (*Arabis blepharophylla*), and the coast fiddleneck, it helps to keep the "rock gardens" attractive through the year.

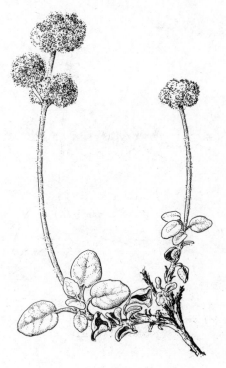

COAST BUCKWHEAT

WESTERN KNOTWEED (*Polygonum bistortoides*)
SMARTWEED (*Polygonum coccineum*)
Buckwheat Family

You may not be fortunate enough to see these one- to three-foot-high plants topped by a thick short spike of small white flowers. They grow in meadowy edges of freshwater lagoons or marshes on the northern part of the Point Reyes National Seashore. It is added to our list of wildflowers for interest as well as beauty. In mountain meadows and boggy streambanks from here to Alaska and in northern coastal marshes, too, the western knotweed grows, but the marshes of Marin County are its south-ernmost limit in coastal California. Several species of smartweed are to be found in and about the sag ponds of the southern part and in ditches and fresh-

WESTERN KNOTWEED

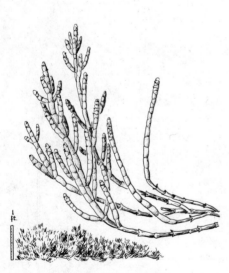

SMARTWEED

water lagoons of the northern part of the National Seashore. One of them (*Polygonum coccineum*) has narrow spikes of rose-colored flowers. Both a floating form and one that grows in the drying mud of the pool margins can be seen. The joints of the leafy stems are swollen, and at each joint is a thin membrane that looks like a short twist of brown tissue paper.

COMMON PICKLEWEED (*Salicornia virginica***)**
Goosefoot Family

As often mentioned, some plants have special requirements for survival such as water, sun or shade, and soil of the places in which they grow. Poison oak and sticky monkey-flower are scarcely indicators of their habitats, for they grow equally well in grassy meadows or forests or in the coastal scrub above the sea. On the other hand, plants like pickleweed or beach morning-glory never stray from their particular habitats: the saltmarshes for the first and beaches and dunes for the second. By their nature they couldn't grow elsewhere. Pickleweed is a strange plant. Along its fat-jointed stems the normal features of a plant— the leaves, flowers, and seeds—are either actually lost or so reduced in size that they can scarcely be seen (except with a hand lens) at the joints of those long green cylinders which are the stems. Perhaps the shape, perhaps the juiciness of these stems give the plant the name "pickleweed." At least they are "pickle" green except in the fall and winter when they add color to the saltmarshes by turning to a greenish purple or even a deep rose.

COMMON PICKLEWEED

YELLOW SAND-VERBENA (*Abronia latifolia*)
Four-o'clock Family

It is not surprising to learn that this sand-verbena and the pink-flowered variety often found growing with it on the sandy beaches, were among the first plants described from the California coast. In fact, the pink sand-verbena (*A. umbellata*) was *the first*, its seeds having been collected in 1786 by members of the ill-fated La Perouse Expedition. With the early exploring expeditions came naturalists or even ships' doctors who collected seeds to enrich European gardens. The area where they could collect the seeds was almost always limited to the seashore. Seeds of the yellow sand-verbena were collected in San Francisco on the Kotzebue Expedition. They were described as a new species by Dr. Eschscholtz later. The fragrant yellow flowers of the sand-verbena can be found in blossom most of the year except in the coldest months. Its sticky foliage is usually coated with sand blown by the perpetual winds on the beach.

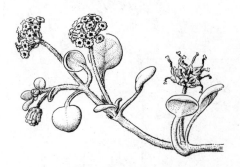

YELLOW SAND-VERBENA

SCARLET PIMPERNEL (*Anagallis arvensis*)
Primrose Family

Scarlet pimpernel is a rather attractive little European weed that has spread from the flower gardens of farms to pastures and grazing land. The pretty little flowers are orange-vermilion with a purple center and—often enough to make the search interesting—a deep blue-flowered form may be found. The capsule is round, really like a small round pot with a lid, and with a delicate touch and fine forceps the top of

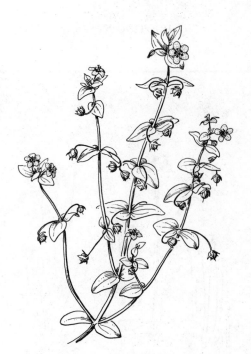

SCARLET PIMPERNEL

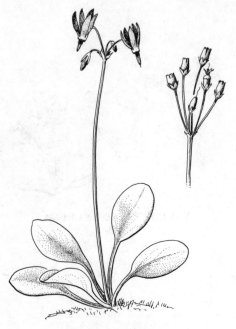

SHOOTING STARS

THRIFT

the "pot" can be picked up by the "handle" (the dried style) to expose a small pot full of seeds in the ripe capsule. Pimpernel has sometimes been called "poor man's weatherglass," for the corolla lobes rewrap themselves bud-fashion when the sky is overcast or when it rains. Pimpernel is under suspicion as a poisonous grazing plant, but authenticated examples of poisoning are hard to come by, according to a recent book on poisonous plants.

SHOOTING STARS or MOSQUITO BELLS (*Dodecatheon hendersonii*) Primrose Family

It is easy to understand that shooting stars are rather closely akin to cyclamens, that favorite potted plant of nurseries and florist shops. The corolla lobes turn backward; the flower has the same pink and purple tones as flowers of the cyclamen; the leaves are all basal and thickish. It is a plant of moist grassy places in brush or open woods. With the milk maids or rainbells (*Dentaria*), shooting stars are among the very first wild flowers to bloom. To those who are fond of them, the objective of walks in the woods in late January and in February is to visit known spots where shooting stars grow to check on the crop of the current year.

THRIFT (*Armeria maritima californica*) SEA-LAVENDER OR MARSH-ROSEMARY (*Limonium californicum*) Leadwort Family

Thrift grows along the seacoast from San Luis Obispo County north to British Columbia. Ocean bluffs, edges of dunes, or even areas just above the line

of the saltmarshes are places appropriate for its growth. A beautiful patch of
the long-stalked bunches of pink flowers grows along Limantour Spit. It is
closely related to the little edging plant
so commonly seen in gardens. It has
the same basal mound of tough grass-
like leaves. The heads of flowers top a
much taller stem than those of the garden plants, but both are pink flowered
and have strawlike bracts and calyces.
They could be used in a dry bouquet.
The plants bloom from late spring
through summer.

Another member of the leadwort
family *(Limonium californicum)* is
found in the saltmarshes growing with
pickleweed and salt grass. Sea-lavender or marsh-rosemary is not as attractive as its name suggests. It is easy to
spot because of its basal clump of
largish leaves. The stem (a foot or so
high) is open-branched, and the purple
flowers are nearly concealed by the
bracts so that you have to look carefully
at the plant to see if it is in flower. The
blooming period is in summer.

SEA-LAVENDER

MANZANITA *(Arctostaphylos)*
Heath Family

The common name means little ap-
ple, and the scientific name means
bear's grape or berry, but in spite of
this advertising the fruits are very poor
eating indeed. The berries, though
somewhat acid tasting, are very dry and
mealy and have large hard seeds. Nev-
ertheless the fruits are pretty to look at
but do not compare in beauty to the
clusters of little white or pinkish tinged
urn-shaped flowers. Other Indian
tribes and probably the local ones
found the berries nourishing, and
Charles Francis Saunders in his book
on western wildflowers tells that white

men as well as red men enjoyed the cider drained from cooked, crushed berries. Many of the western species—and there are many of them—do not have the differences between them sharply defined, and in addition to that it has been shown that natural hybrids occur. Only two of the manzanitas that grow within the Point Reyes National Seashore are described below.

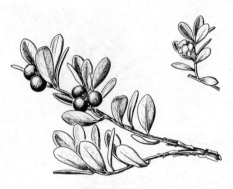

BEARBERRY

BEARBERRY, KINNIKINNIK (*Arctostaphylos uva-ursi*) Heath Family

All the way across the country from the Pacific to the Atlantic this prostrate shrub with creeping woody branches, sometimes tipping upward six to ten inches, grows in suitable cool climates and places. It is small wonder then that there is a lot of variation from coast to coast. One of the several kinds of this creeping manzanita used in the California nursery trade is known as *Arctostaphylos* 'Point Reyes.' The little clusters of white or pinkish flowers are shorter than the leaves. The fruit is bright red and about half an inch in diameter. It is not easy to find because of its infrequent occurrence.

BOLINAS MANZANITA (*Arctostaphylos virgata*) Heath Family

BOLINAS MANZANITA

This tall shrubby manzanita (four to twenty feet high) is as common as bearberry is rare. The smooth chocolate-colored stems are as beautiful in this as in other shrubby species of the genus. The first of the white flower clusters with their slender green basal bracts usually come out by New Year's Day.

The wood-brown berries which ripen
in summer are sticky and so is the hairy
foliage. This species and some other
manzanitas were described by Miss
Alice Eastwood who for so many years
was the botanist at the California Acad-
emy of Sciences. She botanized the
Marin region and in her long association with this fascinating collecting
area discovered much there that was
new and unusual.

SALAL *(Gaultheria shallon)*
Heath Family

Though it prefers to be fog- or rain-
drenched or have its roots in surface or
underground water, salal will grow
anywhere: open pine forests, north-
facing wooded draws, ledum swamps,
or even exposed on Point Reyes itself.
Erect or spreading stems beset with
evergreen leaves make either a ground
cover or a dense thicket which in spring
bear reddish stemmed, reddish bracted
clusters of urn-shaped flowers which
are either white or pinkish. In late summer and fall these flowers ripen into
quarter-inch purplish black berries
that are palatable to birds and beasts.
The early explorers of the west coast
report that the berries were part of the
diet of the northern Indian tribes and
that they, themselves, also found them
worth eating. Like the California huckleberry *(Vaccinium ovatum)*, salal has
been cut in the Christmas season and
shipped to florists for sale. It has been
called lemon-leaf in the trade. It may
be all right to cut plants on one's own
woodlot, but they should be left un-
molested on public lands so that others
can appreciate their beauty.

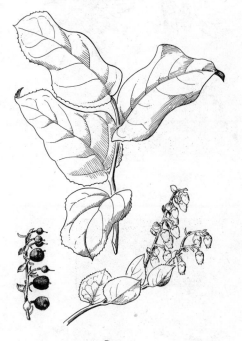

SALAL

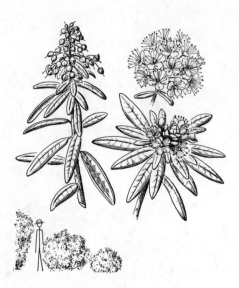

LEDUM

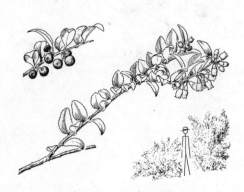

CALIFORNIA HUCKLEBERRY

LEDUM or LABRADOR-TEA *(Ledum glandulosum columbianum)*
Heath Family

This shrub is so characteristic a part of the greenery around seeps and springs on Inverness Ridge that the term "ledum swamp" is used to describe its special habitat. Where you find deerfern *(Blechnum spicant)*, yellow-eyed grass *(Sisyrinchium californicum)*, tinker's penny *(Hypericum anagalloides)*, California bluebell *(Campanula californica)*, the chances are that you will find ledum. Unlike the shrubs of flowering currant and salmonberry that are usually found with it, ledum is an evergreen and does not lose its leaves in winter. The firm-textured leaves are dark green above and whitish beneath and rather aromatic when crushed. It is probably the resinous nature of the leaves in other kinds of ledum that caused them to be used by Indians and early settlers in the east to make tea. The thick clusters of white flowers are easy to see against the dark green of the leaves, so this is still another instance where one can botanize from an automobile. The dry dark brown fruits are also clustered and remain on the shrubs for some time.

CALIFORNIA HUCKLEBERRY *(Vaccinium ovatum)*
Heath Family

The California huckleberry is a handsome evergreen shrub more at home in forest shade where it grows with wax-myrtle and salal *(Myrica californica, Gaultheria shallon)*, than in open places but it can be found in a reduced compact form in the gullies on the ocean bluffs. When it is cut back, new shoots grow readily. The urn-shaped white or pinkish flowers grow on the underside of the branches protected by the flattish sprays of thick, shining, dark green leaves. The beauty of the leaves has been capitalized on by the people who cut the leafy branches (hopefully with the permission of the owner) to sell to the florists. The flowering season extends from April through May. The blue-black huckleberries ripen late in the summer. The berries are just another added attraction of this superb shrub. Though only about one-fourth of an inch long, they are sweet; they are juicy; they are admirable in pies. Cape Cod may have its blue-berried *Vaccinium* immortalized with Cape Cod blueberry grunt but I am sure that a pudding which might be called California huckleberry grunt would be equally tasty.

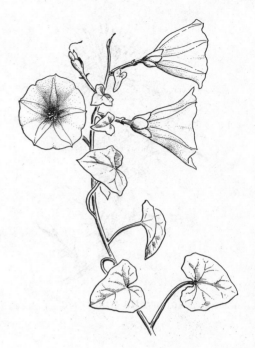

MORNING-GLORY

MORNING-GLORY *(Convolvulus occidentalis)*
BEACH MORNING-GLORY *(Convolvulus soldanella)*
Morning-glory Family

The flowers of the two morning-glories are much the same, that of the beach morning-glory being larger and having the pink color predominant. The obvious differences are in the kinds of places they grow in and the shapes of their leaves.

Beach morning-glory is a true strand plant and will never be found away from the sandy beach or the moving dunes. The leaves are kidney-shaped and quite thick. The other morning-glory may grow in an open field or climb up trees and shrubs or even cover a roadcut with its thinnish angular

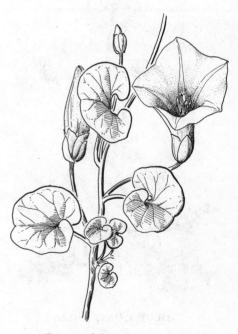

BEACH MORNING-GLORY

Navarretia squarrosa

BLUE COAST GILIA

leaves and white flowers. The outer surface of the white funnel of the flower has five dull pink stripes. When the flowers wither, the pink color becomes conspicuous over the whole flower. Morning-glories from the coastal scrub and headlands may have the white flowers suffused with pink when they open. There is considerable variation in leaf shape as well as color. The leaves are always angular, some much more so than others.

BLUE COAST GILIA *(Gilia capitata chamissonis)*
Phlox Family

Gilias particularly have recently been the basis of intensive genetic and experimental study into the true relationship of species and subspecies. There are many kinds of them throughout the state to furnish material for such a study. Too often in such variable groups too many scientific names have been applied to closely related plants, but after the research that has been done at the Rancho Santa Ana Botanical Garden on the whole phlox family, we can safely say that the annual with heads dense with blue corollas and wooly calyces and leaves much divided into threadlike divisions, found on dunes and sandy flats along the coast from San Mateo County to Mendocino County, can be called *Gilia capitata chamissonis.*

There is another member of the phlox family *(Navarretia squarrosa),* which you will probably smell before you see. It has blue star-shaped flowers that are small. The branching plants which bloom in summer are two to six inches high. The divided needle-pointed leaves are sticky with glandular hairs, and it is these sticky hairs that give forth an odor exactly like a skunk's.

COMMON LINANTHUS (*Linanthus androsaceus*)
LARGE-FLOWERED LINANTHUS
(*Linanthus grandiflorus*)
Phlox Family

Little colonies of the small-flowered linanthus grow in grassy openings among the trees on Inverness Ridge. These small annuals usually have a single wiry stem with a circle or two of leaves which are divided to their bases into stiff needlelike parts. The top circle has few to several corollas with lobes that spread at right angles to their long threadlike corolla tubes. This is a variable species, and in its range up and down the state there are several varieties different enough to have supplementary scientific names. The colors of patches of it you find in the park area vary from bright rose to creamy white, but the color difference is not important enough to rate an official name.

Another linanthus (*Linanthus grandiflorus*) is found only on the stable dunes just back of Point Reyes Beach. It has much the same growth form and leaves as *L. androsaceus*, but the stems can grow to a height of one and a half feet. The flowers are much the same size and shape as the perennial phlox which beautifies home gardens in the summer. The flowers are white or pale lilac.

COMMON LINANTHUS

LARGE-FLOWERED LINANTHUS

YERBA SANTA

BABY-BLUE-EYES

YERBA SANTA (*Eriodictyon californicum*)
Phacelia Family

The tubular lavender-purple flowers of yerba santa can be seen in late May and in June on a colony of the leggy ill-formed shrubs where the road crosses Inverness Ridge on its way to Point Reyes. It is reported that more of it grows toward the Bear Valley and to the southward. The shrubs normally grow in dense stands. They stump-sprout and grow with renewed vigor in response to disturbances like road building and fire. Most of our aromatic native plants have a history telling of their uses by the early settlers—lore that may originally have come from the Indians—and yerba santa is no exception. The firm peachlike leaves, so thickly spread on top with a glutinous resin that they look varnished, furnished a brew that was much used for respiratory ailments. The mission fathers learned of its curative powers from the Indians and named it the holy herb. Whether it was the Indians who also passed on the information that it would pass for both chewing or smoking tobacco I have not been able to discover, but it is known that the early settlers used it this way. West of the high Sierra Nevada from north to south it is a widely distributed shrub.

BABY-BLUE-EYES (*Nemophila menziesii*)
Phacelia Family

Everybody knows the California poppy, and nearly everybody—at least those who tramp the hills in spring—knows baby-blue-eyes, cream cups,

and shooting stars. The baby-blue-eyes
are weak low-sprawling annuals that
bloom almost as early as the milk maids
or rain bells *(Dentaria californica)* that
become so common in February. The
sky-blue, white-centered flowers,
etched with fine lines and dots of blue
so dark that they are nearly black, are
common in open woodlands and grassy
spaces on bluffs above the ocean. Often
the basin-shaped blossoms are bluish
white, but the markings on the corolla
are the same as the bright blue ones.
These have the scientific name of
Nemophila menziesii atomaria. It, too,
has been found at Point Reyes Light-
house and bluffs above Bear Creek, as
well as in other likely places.

CALIFORNIA PHACELIA *(Phacelia californica)*
Phacelia Family

When this stout perennial grows
where it gets the full force of the ocean
wind, it is quite low growing, but when
it is more protected it may reach a
height of one and a half feet. Before the
coiled flower branches expand, the
plants, which are thickly covered with
long straight hairs, have a gray-green
color. As the flowering branches un-
curl, each branch thickly set on the
upper side with long-stamened flowers, the plants present a most subtle
blend of purple and gray-green.

At least two other kinds of phacelias
grow within the park. The one most
likely to be seen is the wild heliotrope
or fern phacelia *(Phacelia distans,* not
illustrated). This is about a six- to ten-
inch annual with leaves divided and
fernlike. The curled flower branches
are not so thickly set with flowers as

CALIFORNIA PHACELIA

FORGET-ME-NOT

COAST FIDDLENECK

the California phacelia. These have dark stamens and a corolla that is white tinged with lavender. This is apt to be found on sandy flats like those around Limantour Spit.

COAST FIDDLENECK (*Amsinckia spectabilis*)
Borage Family

All the kinds of *Amsinckia* have orange or yellow flowers growing on one side of the curving stems which inspire the common name. The different kinds are hard to tell apart, but this one can be identified by the place it grows. It is found on the coastal strand and the upper edges of saltmarshes. At Limantour Spit, for example, it can be seen blooming in spring in dune sand or in rather saline soil with thrift (*Armeria maritima californica*) and other plants that just barely get their feet in the saltmarsh. The trumpet-shaped flowers (about one-half inch long) are brilliant orange yellow and attractive, and the dark hairs on the calyces are effective, but the rest of the plant is not. The leaves and stems are unpleasant to the touch with their covering of coarse bristles which have swollen pustulate bases giving the plant a rather pimply look. The plants vary in habit from completely erect to a sprawling flat mat. This species was first collected on the West Coast, probably at Fort Ross or Bodega and sent to Russia, for it was described by botanists associated with the botanic garden of what was then St. Petersburg.

FORGET-ME-NOT (*Myosotis latifolia*)
Borage Family

The roadsides of the Douglas-fir forests and other shady places are made pleasant by a profusion of light blue forget-me-nots. The calyx with the four ripe seeds inside is lightly covered with delicate hooked hairs. These catch on anything passing by and the seed is spread about. Perhaps this is just a weed as it has been introduced by way of the flower gardens, but like the English daisy, which has much the same history, it is a welcome addition to our spring display.

INDIAN PAINTBRUSH (*Castilleja*)
Figwort Family

Flowers of most of the many species of Indian paintbrushes are bright red. This is true not only of the blossoms themselves but the tips of the calyces and the bracts that all make up the bright-colored spikes of these perennials. However, the most common species found at the National Seashore has pale yellow instead of bright red floral parts. This is *Castilleja wightii* which is most commonly seen close to the coastal bluffs. In brushy areas a bright red species grows. This differs from *C. wightii* in that the corolla itself conspicuously spreads at right angles to the flower spike, and the tubular calyx is much more deeply cut on one side than the other. This is *Castilleja franciscana*. It would be big news for Seashore botany if one other species of Indian paintbrush could be found again. It is *Castilleja leschkeana*, which was collected in the damp swales behind the dunes on Point Reyes Beach and described as new in 1949 by J. T. Howell in his *Marin Flora*, an invaluable reference book about the plants growing in Marin County and the Point Reyes National Seashore.

Castilleja wightii

Castilleja franciscana

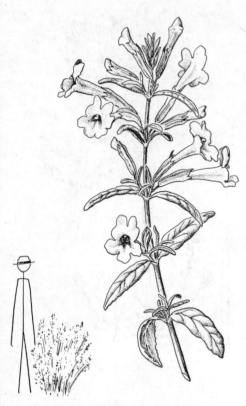

STICKY MONKEY-FLOWER

COAST OWL'S-CLOVER

STICKY MONKEY-FLOWER *(Mimulus aurantiacus)*
Figwort Family

The sticky monkey-flower (see p. 46 also) is a shrub and grows on drier slopes or openings in the bishop pine or Douglas-fir forests or the brushy slopes above the ocean. If you hold a young leaf or even an old one between your thumb and forefinger you will know immediately why this attractive orange-flowered shrub which blooms through late spring and summer is called sticky monkey-flower. It is sometimes believed to be so different that it is placed in another genus *(Diplacus)*, under which name you may find it if you search for more information about the plant.

COAST OWL'S-CLOVER *(Orthocarpus castillejoides)*
JOHNNY-TUCK *(Orthocarpus erianthus roseus)*
OWL'S-CLOVER *(Orthocarpus purpurascens latifolius)*
DWARF OWL'S-CLOVER *(Orthocarpus pusillus)*
VALLEY-TASSELS *(Orthocarpus attenuatus)*
Figwort Family

There are nine species of *Orthocarpus* growing on Point Reyes Peninsula, all of which are spring flowers. Note the difference in the corollas of the five species pictured here. All are drawn to the same scale and show in their flowers alone some of the resemblances and differences that may be found among a few species in one genus. The smallest plant, dwarf owl's-clover *(O. pusillus)*, has completely in-

conspicuous maroon flowers almost
concealed in reddish green foliage.
They are common on grassy slopes and
easy to spot after you become ac-
quainted with their appearance, which
is characterized by the reddish leaves.
The broad flower spikes of coast owl's-
clover *(O. castillejoides)* are conspicuous, and both the bracts and flowers
are colored; the bracts are usually a
pinkish white; the flowers buried
among them are yellow with deep purple markings. Patches of these plants
can be seen on the borders of saltmarshes from Monterey County northward. It is common at Limantour Spit.
Valley-tassels *(O. attenuatus),* is an
erect plant rising stiffly among the grass
of the hillside. Even the white-flowered spike is narrow and straight. This
is another instance of a species native
to two continents. It grows in Chile and
the coastal area of western North
America. The prettiest of the five is the
white form of Johnny-tuck *(O. erian-
thus roseus).* The three conspicuous
puffy white sacs of the lower lip turn
pink as the flower ages. The sharp upper lip is like a straight purple spine.
The blending of these colors, white,
rose, and purple in the dense colonies
of these small annuals, is a delight to
the eye and to the nose, too, for they
have a delicate fragrance. *Orthocarpus
purpurascens latifolius* is a form of the
common owl's-clover that often paints
interior fields and slopes with rich rose
purple. This form is found on the dunes
and coastal bluffs and differs in having
the flower bracts much more divided
and often a lighter pink giving the inflorescence a banded appearance. The
plant is fairly common on Point Reyes
Beach.

JOHNNY-TUCK

OWL'S-CLOVER

DWARF OWL'S-CLOVER

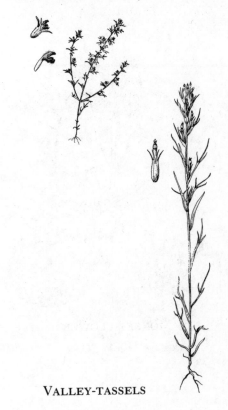

VALLEY-TASSELS

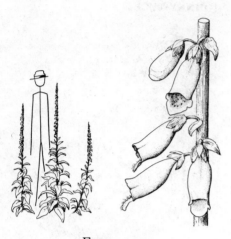

FOXGLOVE

MONKEY-FLOWER

FOXGLOVE *(Digitalis purpurea)*
Figwort Family

This is another plant of Point Reyes that can be identified from a moving automobile. Many people will recognize these large white- or rose-colored pendent flowers as the familiar foxglove of the garden and they will know that it has now become a natural part of the floral landscape since it grows where no one could possibly have planted it. It is a biennial and reseeds itself the second year. The old English name "foxglove" is said to be derived from "foxes glew," an old Saxon musical instrument made up of bells hanging on a frame. The association of drooping flowers and hanging bells is quite obvious. Its early use as an old household remedy for heart disease foreshadowed its present great value in treatment of ailments of the heart.

MONKEY-FLOWER *(Mimulus guttatus)*
Figwort Family

The bright yellow monkey-flowers speckled with reddish brown spots remind one more of snapdragons than faces, either those of monkeys or humans, in spite of the meaning of the scientific name. The almost succulent herbs are always found in wet places: roadside ditches, springs, pond edges and seeps at bases of the cliffs above the beach. It is a plant showing great variation in form. The form most often seen in the National Seashore is a lush large-flowered one with the upper part of the stem and the leaves somewhat fuzzy.

COYOTE-MINT *(Monardella villosa franciscana)*
Mint Family

Clumps of leafy stems about one foot high, with herbage giving off a strong mint odor if you brush against it, tells you that you have found the perennial coyote-mint. The leaves which are opposite, like all leaves in the mint family, have the veins made prominent by being impressed on the upper surface and rather ridged on the under leaf surface which is covered by short white hairs. The leaves and bracts that are just below the densely flowered heads of small purplish pink blossoms are beset with the same type of short velvety hair. This perennial is not abundant and yet not rare in the coastal scrub or even on open rocky spots.

COYOTE-MINT

WESTERN PENNYROYAL *(Monardella undulata)*
Mint Family

There is an annual species of coyote-mint or western pennyroyal *(Monardella undulata)* that grows on dunes and sandy flats of Point Reyes Beach. It is not a common plant elsewhere, as it grows only along the coast from Santa Barbara County to Marin County. The branching plants are a foot or less high and bear several heads of royal purple flowers. They grow erect from the long annual roots buried deep in the loose sand of the dunes. It is stunning when seen growing with the large-flowered white linanthus. A dune slope covered with these two plants would be irresistible to anyone with color film in his camera.

WESTERN PENNYROYAL

SELFHEAL

SELFHEAL *(Prunella vulgaris atropurpurea)*
Mint Family

Selfheal grows on grassy hillslopes, mostly those facing the sea. It may grow fifteen inches high, but more commonly is much less when it is a part of the windswept turf of the headlands. The wiry creeping rootstock helps to spread it. The dark-green, hairy foliage lacks the tangy odor one associates with mints. The flowers are clustered into a short dense spike. They are two-lipped, of a rich violet color, and partly concealed by a broad, veiny purplish bract. This form of selfheal is a native of the northern hemisphere. A white-flowered form is sometimes seen.

HEDGE-NETTLE

HEDGE-NETTLE *(Stachys rigida quercetorum)*
Mint Family

COAST HEDGE-NETTLE *(Stachys chamissonis)*

The growth of the coast hedge-nettle is rank, the odor is rank, but the spike of inch-long rose-purple flowers that top the square erect stem makes up for such shortcomings. Only in swampy areas thick with grasses and sedges or moisture-loving shrubs will you find coast hedge-nettle. It blooms in summer.

Its close relative *(Stachys rigida quercetorum)* will grow almost anywhere except in the saltmarshes or in full hot sun. It really prefers a shaded north slope. The plants are weak-stemmed. The whorls of the flower spikes are made up of shorter and paler pink blossoms that have much the same shape as coast hedge-nettle. The blooming period begins in spring and continues in favorable localities well into summer. Why "nettle" is part of the common name is not clear, because hairs on the leaves do not sting. The name came to us from Great Britain, where the native species may perhaps look more like the common annual nettle.

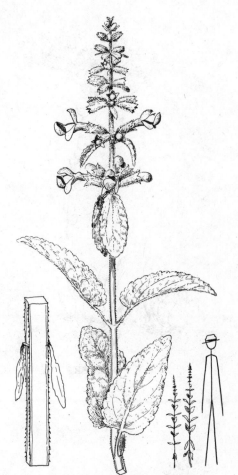

COAST HEDGE-NETTLE

YERBA BUENA *(Satureja douglasii)*
Mint Family

The aromatic odor that rises when one crushes one of the long creeping strands of yerba buena is pleasant indeed. No wonder the early settlers brewed a tea of it. The tea was supposed to cure many things, but to have the brew taste good was the surprise. Cooling summer drinks can have their flavor enhanced by the addition of the "good herb." The small (about one-fourth inch) tubular white corollas grow in the axils of the leaves and are not particularly noticeable, and the small dry nutlets are even less so. The plant grows in forest and shrubbery and sometimes ventures out to the open hillsides.

YERBA BUENA

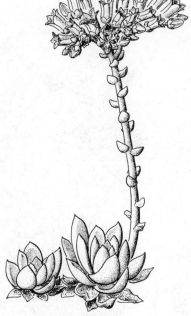

LIVE-FOREVER

LIVE-FOREVER *(Dudleya farinosa)*
Stonecrop Family

These succulent plants of sea bluffs and rocks are just as frequently known by the name "hen-and-chickens" as "live-forever." Two or three other genera that grow in home rock gardens, while still not in bloom, look enough like this to be brothers, at least, of *Dudleya.* The juicy triangular leaves that grow in a basal rosette are bright green or else lightly overlaid with a chalky dusting or may even show a distinctly reddish hue. The fleshy flowering stem branches at the top and bears a more or less flattish cluster of pale yellow flowers.

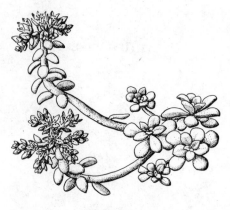

SEDUM

SEDUM or STONECROP *(Sedum spathulifolium)*
Stonecrop Family

People who grow succulents know there are many kinds of sedums and that there is a lot of variety in leaves and growth form. The sedums of Point Reyes grow in thin soil around rock outcrops. The blunt bluish-green leaves grow in a rosette and some straight leafless stems radiate from the rootcrown, each one of them terminated by a smaller flat rosette of leaves. The yellow flowers are borne on a two- or three-inch stem that is sparsely beset with small fat leaves. This species grows in the coastal mountains and the Sierra Nevada as far north as British Columbia.

BROOK FOAM *(Boykinia elata)*
Saxifrage Family

The leaves look something like the
alum-roots but are lobed more deeply
with the lobes sharp-toothed. They
look quite a bit like cut-leaved maple
leaves. Brook foam is a water-loving
perennial and grows along living
streams or the surface water in ledum
swamps, blooming from June to Au-
gust. The loose airy clusters of five-
petaled white flowers are beautiful
while they last. You would not pick
them anyway, but if you were to do so
the petals would drop long before you
reached home.

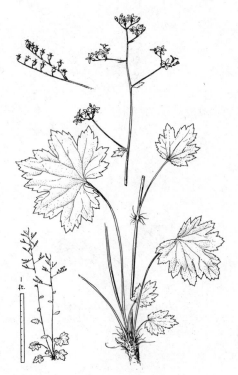

BROOK FOAM

ALUM-ROOT *(Heuchera micrantha)*
ALUM-ROOT *(Heuchera pilosissima)*
Saxifrage Family

If only the lowest leaves of the two
species of alum-root and of fringe-cup
and of sugar scoop *(Tiarella unifoliata)*,
too — if you could find it — were placed
side by side, even the botanists might
be confused by their similarity. The
plants themselves — their way of
growth, their fruits, their flowers — present a problem to no one.

The alum-roots, according to the
books, have a puckery taste to the long-
ish thickened rootstock from which the
flowers and leaves grow; hence the
name. The leaves of *Heuchera micrantha* are frequently reddish veined, and
the great plumes of tiny flowers, one to
three feet long, are frequently tinged
with pink, except for the stamens and
style. It grows commonly on shaded
banks or cliff faces or even north-facing
roadcuts. There is a good floral display
of it in early summer along Sir Francis

Heuchera micrantha

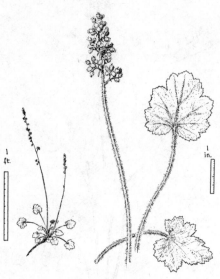

Heuchera pilosissima

CANYON GOOSEBERRY

Drake Boulevard beside Tomales Bay.
Heuchera pilosissima is to be expected
on the coastal bluffs and gullies of the
headlands. The leaves are like those of
H. micrantha, but the calyx is densely
hairy and the inflorescence is not open
and feathery but much more compact
with shorter branches. The flowers of
both are similar.

CANYON GOOSEBERRY (*Ribes
menziesii*)
Saxifrage Family

In the genus *Ribes*, according to
Munz's *California Flora*, one finds all
the species of the currants and of the
gooseberries under the one name. In
other botany books the gooseberries
are sometimes found under the name
Grossularia. There are at least five
kinds of gooseberries in Marin County,
some of which are considered to be
simply varieties of the canyon gooseberry. These thorny, often single-
stemmed, widely branching shrubs are
found on shady slopes or in the deep
wooded bluffs above the sea. As Miss
Parsons has said in her *Wild Flowers of
California*, the hanging flowers remind
one of little fuchsias. The petals are
white. The dull red-purple sepals turn
back at flowering and later dry in a
forward-pointing position, remaining
on the ripening berry which is covered
with small but sharp spines. Wild
gooseberries have been used in making
jelly, but it is a laborious process.
Nearly all western species are spine-
covered to some degree.

FLOWERING CURRANT (*Ribes sanguineum glutinosum*)
Saxifrage Family

While it is still winter the fan-folded leaves of currant begin to open, and the pink flowers on the nodding flowering stalk begin to expand. The plants grow here and there, usually on north-facing slopes. They attain their best growth in ledum swamps, and some bushes are at least fifteen feet high. Leaves and even fruit exude a sticky substance that emits a tangy aroma which many find pleasing. The black, often bloom-covered currants ripen in summer. They generally taste flat and insipid. This is one of the plants that has been enjoyed in English gardens for over 125 years principally through the efforts of David Douglas who was sent out to collect plants by the Royal Horticultural Society in the early part of the nineteenth century. Baby-blue-eyes, cream cups, and California poppies were well known abroad before American gardeners on the eastern seaboard knew of them.

FLOWERING CURRANT

FRINGE-CUP (*Tellima grandiflora*)
Saxifrage Family

Fringe-cups (*Tellima grandiflora*) have clustered straight stems about two feet high with only a few scattered stalkless stem leaves. The green cups (calyces) are set closely along the upper stem. The petals, inconspicuous but worth looking at closely, are pinkish red to red with margins deeply fringed and curled back against the cup. Plants are found near streams in shaded forests and thickets, and are quite common on Bear Valley trails.

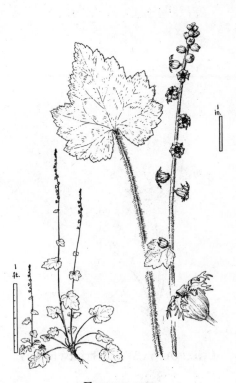

FRINGE-CUP

BEACH STRAWBERRY

CALIFORNIA WILD STRAWBERRY

BEACH STRAWBERRY (*Fragaria chiloensis*)
Rose Family

The scientific name of the species refers to the place where these plants were originally collected—the island of Chiloe off the southern coast of Chile. All the way up the Pacific Coast, skipping the tropics of course, as far north as Alaska the firm shining evergreen leaves, large white flowers, and red strawberries along with other low-growing plants help make the turf of the ocean bluffs a thing of beauty. It will grow in the dunes also. The flowers that have pistils and embryonic seeds and the flowers with stamens only are more often on separate individual plants; that is to say that both sexes are not in the same blossom as is so often the case with plants. Therefore, the flowers that have only stamens will not set fruit. The large (over 1/2-inch in diameter) berries have a delicious flavor and have been used in breeding some cultivated strains. In recent years a hybridized relative has come into use as a garden ground cover and provides an extra dividend of tasty fruits.

CALIFORNIA WILD STRAWBERRY (*Fragaria californica*)
Rose Family

Strawberries of the woods are much smaller than the beach species, and if the weather is not right may have no juice at all, but a fragrant juicy berry is delectable. The three leaflets do not look much different at first glance from those of your own strawberry patch. Everyone can recognize strawberry

plants. Wild strawberry leaves are thin
and sparsely covered on the underside
with flat silvery hairs. The five-petaled
white flowers are usually less than 1/2-
inch wide.

CREAM BUSH or OCEAN SPRAY
(*Holodiscus discolor franciscanus*)
Rose Family

In a mixed woodland of trees and
shrubs the presence of individual
plants is not noticed as you drive along
the country roads unless the plants are
in bloom. Ocean spray is not as compel-
ling a sight as buckeyes. It is not a
small tree but rather a not very abundant shrub with several stems which
are five to fifteen feet high. It is reported that northern Indian tribes
found the straight stalks good for ar-
row shafts. The toothed leaves vary
from one to three inches in length and
look sturdy enough to stay on the twigs
all year, but even along the seacoast
with its more even climate they drop
in the winter. The plant comes to its
glory in midsummer when foot-long
(or shorter) branching sprays are covered with a mass of small cream-
colored blossoms. This certainly is a
shrub that nurserymen might well
propagate and sell to us for our gardens.

CREAM BUSH

OSOBERRY (*Osmaronia cerasiformis*)
Rose Family

If you discover this when the almost-
black fruits are hanging from the
branches you will think you have found
some kind of plum. You will at least
have guessed the right family. Osoberry has yellowish green leaves that
fall in winter. The hanging branches of

OSOBERRY

greenish white flowers with stamens are on one plant; the flowers that will form the fruits are on another. The botanists call this a dioecious plant, that is, one having the sexes on different plants. The pulp of the blackish oblong fruit is very thin and tastes bitter. The new leaves come early in spring, and the odorous flower clusters arrive shortly after. They grow to a height of three to fifteen feet in damp places mixed with other shrubbery. The plant is frequently seen in gullies along the ocean bluffs or in marshy areas.

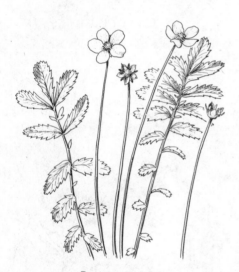

SILVERWEED

SILVERWEED or CINQUEFOIL
(Potentilla egedii grandis)
Rose Family

There are many species of *Potentilla* from the seacoast to the highest mountains, and many of them deserve the common name "cinquefoil" more than this one, for the leaflets of the cinquefoils, often only five, spread out from the end of the leafstalk like lupine or buckeye leaves. With silverweed the leaflets are opposite each other along the midrib of the leaf in a featherlike arrangement. The whole leaf with the many toothed leaflets may be around ten inches or even double that in length. They are silvery white on the underside and green above and grow erect from the creeping stems which root easily and spread the plant about. The flowers grow singly from the leaf axils and look much like buttercups and are just as yellow. They grow in wet places on the coastal strand or lagoons or even at the edge of the salt-marshes.

THIMBLEBERRY *(Rubus parviflorus)*
SALMONBERRY *(Rubus spectabilis*
franciscanus)
BLACKBERRY *(Rubus ursinus)*
Rose Family

THIMBLEBERRY

The kitchen-middens with their
bones and shells can show where the
local Miwok tribe of Indians got their
proteins, but evidence of the rest of
their diet has been long since de-
stroyed. Not all the seeds and berries
of the freshwater marshes and their
surrounding slopes would have been
edible or even safe for them. Salmon-
berries, mild-flavored fruits, are sel-
dom found whole, for the birds like
them too well. The plants with their
arching canes make shrubby tangles in
swamps and swales. They belong to
the rose family, that plant family which
offers us so much food: strawberries
and other berries, plums, apples,
peaches, almonds, and so forth. The
flowers of the thimbleberry spread
rather flatly and the petals almost over-
lap instead of being semi-erect as those
of the salmonberry, and the petals of
the thimbleberry are white instead of
rose colored. The white petals of the
native blackberry lie flat and do not
touch each other but the plant is un-
mistakable. Its trailing leafy stems are
likely to trip you up and snag you with
thorns in almost any grassy or partially
wooded area in the Point Reyes National Seashore. It was first collected by
the naturalists on the "Rurick" when
Captain Kotzebue initially visited San
Francisco. Some flowers have stamens
only, and consequently do not set fruit.
The Himalaya berry *(Rubus procerus)*
usually with five leaflets per leaf, has
escaped to the roadsides and waste
places.

SALMONBERRY

BLACKBERRY

BRINE MILKVETCH (*Astragalus pycnostachyus*)
Pea Family

This is a true saltmarsh plant and will hold its own with its companion plants that also need saline conditions to be at their best. The erect stems are hollow and reddish. The yellowish or dirty white flowers, blooming in summer, are set on the spikes and give a shingled effect, each one curved downward and touching the one below. All the flowers seem to set seed, and the dense spikes of small papery pods, each with a sharp point or beak, are more attractive than the flowers. This is another plant collected by Henry Bolander in the 1860's while he was making plant collections under the auspices of the Geological Survey of California.

BRINE MILKVETCH

FRENCH BROOM (*Cytisus monspessulanus*)
SCOTCH BROOM (*Cytisus scoparius*)
Pea Family

These are two European garden shrubs so well adapted to the whole central California coastal climate that they have taken to the hills to make bright streaks of yellow on hillslopes adjacent to towns and along roadsides. The Scotch broom has angled, nearly leafless stems with largish yellow flowers solitary in the axils of the leaves or, at least, the place where the leaf has been. The seeds in the brownish black pods germinate very readily as do those

FRENCH BROOM

of the French broom. The flowers of
the French broom which are smaller
than those of the Scotch broom are in
small tight clusters along the very
leafy stems. Both push their way into
the landscape and add much color to
the hills in the spring months.

SCOTCH BROOM

BEACH PEA *(Lathyrus littoralis)*
Pea Family

The sweet peas of the flower gardens
belong to the genus *Lathyrus*. So do
both plants pictured here. The beach
pea when in full bloom in midspring is
to me the most beautiful of the beach
and dune plants. Even after the lav-
ender-bannered, white-winged flowers
have passed, the mats of silvery white
foliage are still pleasant to see. The
leaves of the beach pea have no tendrils.

BEACH PEA

WILD PEA

BIRD'S-FOOT TREFOIL

WILD PEA *(Lathyrus vestitus puberulus)*
Pea Family

It is the tendrils on the leaves of *L. vestitus puberulus* that give support to the plants and bring the clusters of lavender-pink flowers to eye level. The flowers turn to a buckskin shade as they age. The plants are not abundant, but may sometimes be seen climbing mountain lilac *(Ceanothus)*, coffee-berry *(Rhamnus)*, or other shrubs in clearings in the forested slopes at the southern end of the Point Reyes Seashore.

BIRD'S-FOOT TREFOIL or LOTUS *(Lotus corniculatus)*
Pea Family

This is a European plant originally introduced for forage that, like many other European herbs, has found the California climate so congenial that it is quite at home growing with the native species. The flower clusters are bright yellow; the leaflets bright green and smooth; the pods—straight and slender and about an inch long and opening when ripe to spill out the seeds—are as many as the flowers and clustered like them. It has become quite common within the park area and is often seen on roadsides and grass-covered grazing land.

HAIRY LOTUS *(Lotus eriophorus)*
Pea Family

The fruit (pods) of this lotus are very different from those of the other two species mentioned here. The much smaller pods are usually one-seeded and have a slender angular beak, looking altogether not unlike an ill-used hand scythe. The pods do not open readily, usually not at all. The perennials are mat forming. A healthy plant may make a mat two feet across, and all parts of the plant except the petals are covered with soft spreading hairs. The loose heads of tiny yellow flowers are in the leaf axils and are shorter than the whole leaf, actually not much wider than each leaflet is long. The flowers take on a reddish cast as they fade. Often growing with this is another flattened lotus plant that differs by having no hairs (or very obscure ones). It is called *Lotus junceus biolettii*. Both are plants of forest land or open slopes near the forest.

HAIRY LOTUS

COAST LOTUS or TREFOIL (*Lotus formosissimus*)
Pea Family

If the plants in this book were grouped by color, the spring-blooming coast lotus would have to be put in two places, for the color of each flower is equally divided between bright yellow and bright pink. The banner—the erect petal in any flower like the sweet pea—is yellow. The wings and keel—the flattened side petals and folded middle one—are pink. Understandably the species name means most beautiful. The pods are slender, straight, about an inch long and open when ripe. The plants are low growing and spread from their perennial roots. They are found in damp swales and on grassy headlands and are native up and down the coast from Monterey County into Washington.

COAST LOTUS

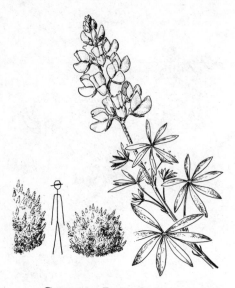

COASTAL BUSH LUPINE

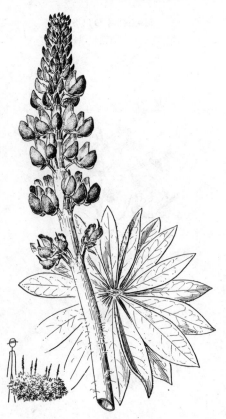

BOG LUPINE

COASTAL BUSH LUPINE *(Lupinus arboreus)*
ANNUAL LUPINE *(Lupinus bicolor)*
BOG LUPINE *(Lupinus polyphyllus)*
VARICOLORED LUPINE *(Lupinus variicolor)*
Pea Family

All over California, mountains and deserts alike, there are many species of lupines, and added to this host of species are variations and some natural hybrids. The bees seek these fragrant flowers and add their bit to the mixing up of species. Many kinds, even annuals, grow at the Point Reyes National Seashore. The shrubby ones along the dunes and the hills above them present a variety of color at the height of the blooming period in May. The most common shrub *(L. arboreus)* is usually yellow flowered but may also be found with flowers that range from lavender to white. Variations in hairiness and smoothness of the stems and leaves as well as in color lead one to think that there has been some mixing by the insect pollinators with some of the matforming shore lupines that grow beside them to explain this difference in appearance. The prostrate lupines on the dunes, bluffs, and headlands are closely related to each other and differ only in some technical points. One, Layne's lupine *(L. layneae)*, grows only

on Point Reyes Peninsula. The one illustrated is *Lupinus variicolor,* which
grows from San Luis Obispo County to
Humboldt County. Its short cluster of
purplish-blue and white flowers rise
from the flat leafy stems scarcely higher
than the bright blue- and white-flowered annual lupines that grow on the
banks with it. The two species of annual lupines in the National Seashore
are just about as variable as their perennial relatives. The one pictured is
a form of *Lupinus bicolor.* The other
species (*L. nanus,* not illustrated) is
called, at least in its larger flowered
form, the sky lupine. Hundreds of these
clusters of whorled flowers paint the
hillslopes in spring. They often grow
with California poppies. But one is
never confused by variation in the bog
lupine (*L. polyphyllus*) in the Point
Reyes region, as there is but one kind,
the tall hollow-stemmed, leafy plants
that grow in the soggy marshes with
other bog-loving plants. The flowers,
many in thick spikes, are more nearly
purple than those of other local lupines.

ANNUAL LUPINE

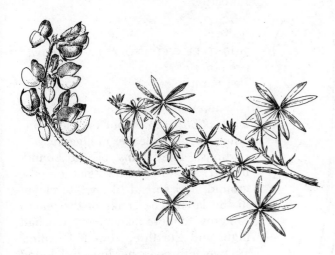

VARICOLORED LUPINE

SPOTTED BUR-CLOVER

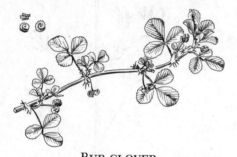

BUR-CLOVER

ROUND-LEAVED PSORALEA

SPOTTED BUR-CLOVER *(Medicago arabica)*
BUR-CLOVER *(Medicago polymorpha)*
Pea Family

The bur-clovers are typical of the association referred to earlier as "farm-side roads and pastures," and like practically all the typical assemblage of plants found there, they are European weeds that have become naturalized. Along with filaree *(Erodium)* it is a "good" weed. All the species of bur-clover are widely diffused in the state and are recognized by stockmen to be excellent forage whether in the fresh and green or in the dry state. In our area it has spread from pasture to the grazing land of the grassy hills and headlands, and over the years these have become covered with just about an equal proportion of introduced and native species. The tiny bunches of yellow flowers are inconspicuous. The bur is a spirally coiled pod with the ridged edge beset with hooked or curved prickles. In spite of the prickles the burs are reported to be good food for cattle.

ROUND-LEAVED PSORALEA *(Psoralea orbicularis)*
Pea Family

Wherever cow clover *(Trifolium wormskioldii)* grows one may see some other cloverlike leaves of gigantic proportions. The stems of the round-leaved psoralea grow flat on the ground, and spread like white clover in the lawn. The stalks of the leaves can grow to more than a foot and a half high, and the three rounded leaflets are two or three inches long and dotted with dark glands which give the plant its characteristic odor. The leafless stalks that bear the dense oblong clusters of flowers rise above the evenly spaced leaves. The small violet-colored flowers appear even a darker shade because of the black hairs on the calyx. According to the botany books, the species grows in cismontane California, which is to say that they are found only west of the mountains (the Sierra Nevada) on the Pacific drainage.

Trifolium wormskioldii

CLOVER *(Trifolium)*
Pea Family

The few rather modest annual clovers illustrated were chosen for a purpose. The large red-purple headed one, the cow clover *(Trifolium wormskioldii)*, which grows in wet places throughout the west, would equal two heads of each of the others pictured which, though charming, are small. On most any grazed-over, open hillside along with attractive and unattractive weeds such as windmill pink *(Silene gallica)*, the cats' ears both the hairy and the smooth *(Hypochoeris radicata* and *H. glabra)*, filarees (Erodium), and a host of European grasses, these natives have been able to hold their own successfully. The clover with one or two heads set closely against leaves *(T. macraei)* and the puffy, red-purple flowered *(T. depauperatum)* have the distinction of growing *as natives* on two continents. These, some other clovers, some lupines closely related to ours, another variety of our red maids

Trifolium macraei

Trifolium depauperatum

Trifolium microcephalum

(Calandrinia ciliata menziesii), and several other plants that are first cousins or identical twins, occur on the coast of Chile and the coast of California. The climates are much alike, but that does not explain everything. How plants get around the globe is a fascinating subject, and not all the answers have been worked out.

The hairy clover *(T. microcephalum)* has pinkish white flowers in the heads that are set in a toothed saucerlike collar. *Trifolium gracilentum*, with reddish-purple flowers, has no collar. The flower heads lack this circular bract which is so characteristic of many of the clovers, and as it develops seeds, the drying flowers are reflexed without this support. For no reason that I can discover the clover with the slightly larger head is called tomcat clover *(T. tridentatum)*. It has reddish-purple flowers tipped with white. There are thirty-odd species of clover in Marin County. It would be interesting to find how many kinds grow within the Point Reyes National Seashore.

Trifolium gracilentum

Trifolium tridentatum

COMMON WILD VETCH *(Vicia americana)*
GIANT VETCH *(Vicia gigantea)*
Pea Family

If in your hikes you observe flowers of vetch *(V. americana)* and certain kinds of the wild sweet pea *(Lathryus)* in the same day, you may begin to wonder why they are not in the same genus. They are closely related, and various technical points separate them. One tiny but marked difference is amusing to demonstrate—if you have a hand lens. The style (slender elongation of

COMMON WILD VETCH

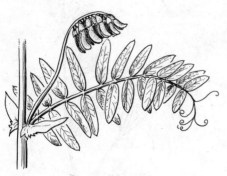

GIANT VETCH

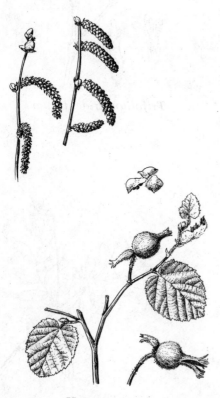

HAZELNUT

what will be the seed pod) is terminated by a microscopic "shaving brush" in the vetches, while in the sweet pea flowers the microscopic hairs at and near the end of the style are only on one side of it, rather like a floor-brush. The lavender-colored flowers are on low-growing plants often in open spaces in the woods.

The giant vetch *(V. gigantea)* is not too attractive an herb, but its abundant growth over the shrubs of the moist coastal scrub commands attention. The flowers are almost as abundant as the foliage, and the dense one-sided racemes of flowers are a rather unattractive maroon-red color. The ripe pods are black.

HAZELNUT *(Corylus cornuta californica)*
Birch Family

The spreading hazelnut bushes lose their leaves in late fall and early winter. By February or earlier the catkins that hold the pollen appear and hang from the tips of last year's branches. If the branching style and stigma (pistil) atop the ovary were not bright red the female flowers on the bushes would scarcely be noticed. The hairy-toothed leaves are reported to be good browse, and the ripe hazelnuts inclosed in their husks disappear into the food caches of squirrels, wood rats, or any other kind of wild life that likes them. If you should be fortunate enough to find a ripe nut you will see that it is a small replica of the filbert or hazelnut of commerce. Bushes are common in any of the forests, either of bishop pine, Douglas-fir, bay and live oak and buckeye, or even in the but rarely found thickets of chinquapin on Inverness Ridge.

CALIFORNIA WAX-MYRTLE (*Myrica californica*)
Wax-myrtle Family

All along the immediate coast from Los Angeles County to the state of Washington the wax-myrtle can be found on most hillslopes or in canyons, and it is quite common on the Point Reyes peninsula. The chances are that the dark evergreen shrubby bush or even small tree that at first one thinks is an unnatural looking bay tree will prove to be a California wax-myrtle instead, but even though fruits or flowers may not be present, a sniff of a bruised leaf will tell you which one it is. The male and female flowers are on the same shrub. The clusters of small dark nutlike fruits are lightly coated with wax but certainly not in sufficient quantity to make bayberry candles. The eastern species (*Myrica cerifera*) is used for that purpose.

CALIFORNIA WAX-MYRTLE

BEACH-PRIMROSE (*Camissonia cheiranthifolia*)
Evening-primrose Family

On the coastal strand from southern Oregon to Baja California, beach primrose can be found in dunes and beaches, with those south of Santa Barbara being larger flowered and more woody. The leafy branches, flat in the sand, radiate from a central rosette of leaves. The shallowly cupped yellow flowers, about 1/2-inch wide, grow in the leaf axils toward the tips of the branches and bloom for a long period in spring and summer. In most local floras this species is listed under the genus *Oenothera*. The genus *Camissonia* and its many western species can be recognized quickly by the ball-

BEACH PRIMROSE

SUN CUPS

GODETIA

shaped instead of four-parted stigma in the flowers and the fact that the flowers open in the morning rather than in the evening.

SUN CUPS *(Camissonia ovata)*
Evening-primrose Family

The yellow flowers of the sun cups appear in spring on open grassy hills and headlands that have not been overgrazed. The plant's only leaves are in a basal rosette that hugs the ground, and the root is perennial so its chances of survival are good. The four-petaled flowers are about an inch across, and what appears to be a flower stalk is part of the flower itself. The hard-ridged capsule, which is the lower part of the flower, is hidden among the basal leaves. In most botany books you will find sun cups listed under the name *Oenothera ovata,* but these are changing times so let us learn the scientific name listed at the head of this paragraph.

GODETIA or FAREWELL-TO-SPRING
(Clarkia amoena)
Evening-primrose Family

This species of *Clarkia* has been known at times by the romantic name of "summer's darling," and this is what it seems to be, coloring the dunes and bluffs along Point Reyes Beach. This particular strain of *Clarkia amoena* grows almost within the range of salt spray blown from the waves. The plants are sprawling, and at the height of the season, loaded with lovely lavender-pink cup-shaped blossoms with each petal at least an inch and a half long, and blotched *centrally* with red. At a distance from the shore the plants may be taller (up to two feet) and the flowers a bit smaller. The seeds sold under the name godetia, the generic name for this group until extensive research was done, were developed horticulturally from the big-flowered coastal form that brightens the dunes in summer. Among the three or four other kinds of *Clarkia* found around Marin County one that looks much like *C. amoena* may even grow near it. It is *Clarkia rubicunda*, and the most obvious difference between the two is the position of the red blotch which in *C. rubicunda* is at the base of each petal—never in the middle of the petal.

CALIFORNIA LILAC (*Ceanothus*)
Buckthorn Family

California lilac comprises the largest group of native ornamental shrubs in the state except possibly the manzanitas. Botanists are not always in perfect accord about what constitutes a species so the number may vary. In Munz's *California Flora* 43 species are listed, not counting subspecies and varieties, under both *Ceanothus* and *Arctostaphylos*. The masses of blue, or in some cases white, clusters of flowers of California lilac have delighted people for years and the plants are sufficiently conspicuous to be appreciated while whizzing by in an automobile. It is small wonder that many kinds have been welcomed into gardens. The horticultural possibilities of many kinds have only recently been exploited in the West, though certain ones have been grown and hybridized in Europe for a long time.

The flowers of *Ceanothus* can produce a soapy lather when rubbed on

BLUE BLOSSOM

wet hands but if the little stems are left on each flower your soap may be greenish.

POINT REYES CREEPER *(Ceanothus gloriosus)*
GLORY MAT *(Ceanothus gloriosus porrectus)*
BLUE BLOSSOM or BLUE BUSH *(Ceanothus thyrsiflorus)*
Buckthorn Family

POINT REYES CREEPER

Blue blossom is quite common on Point Reyes Peninsula and in most of west central California as well. When it grows with other shrubbery it may be 10 or 15 feet high, but much less in the open where the branches spread angularly. At Point Reyes and along Point Reyes Beach it can be forced to creep along the bluffs by the force of the ocean winds. Some shaping of the shrubs can be seen away from the coastal bluffs for deer seek it for browse and plants are occasionally cropped into neat pyramids of green leaves. This plant was collected in San Francisco in 1816 during the visit of the *Rurick* and later described as new by one of the naturalists of its expedition, Dr. Eschscholtz.

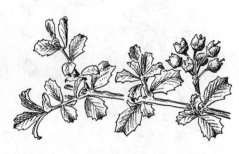

GLORY MAT

The abundant small clusters of blossoms on *C. gloriosus* and its variety *porrectus* are lavender-blue instead of a pure blue shade like those of the blue blossom. The leaves are thick and holly-toothed. Those of variety *porrectus* are often less than one-half inch long, and the margin may be scarcely toothed. This form is found only in the bishop pine forest of this area and grows completely prostrate in sunny openings in the forest. *Ceanothus gloriosus* is nearly prostrate but tends to mound. Deer also prune the ascending branches of this species.

POISON HEMLOCK (*Conium maculatum*)
Parsley Family

Poison hemlock can be considered to be a beautiful annual if one is completely unprejudiced. Its slender tree-like form, sometimes nine feet in height, delicate fernlike foliage, and the flat clusters of small white flowers would be well worth seeing if there were not so much of it. It is common along every country road, particularly near farms and slopes that have been cleared of trees and brush, and it is poisonous! It has been known to be poisonous since classical times. A brew from the root was given to Socrates after he was sentenced to death, so history tells us. Fortunately it cannot be confused with any of the edible members of the family, such as carrots or celery.

POISON HEMLOCK

COW-PARSNIP (*Heracleum lanatum*)
Parsley Family

If the color pictures you have taken in the National Seashore are for scenic effects and include vistas of coastal headlands, or valley streams that are partly shaded and partly in the sun, someone will undoubtedly inquire, "What is that white thing I see scattered around the landscape?" It will be cow-parsnip. The plant is a heavy, one-stemmed herb that grows from about four to ten feet in height. The leaves

COW-PARSNIP

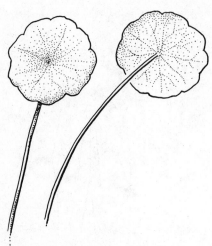

Hydrocotyle verticillata

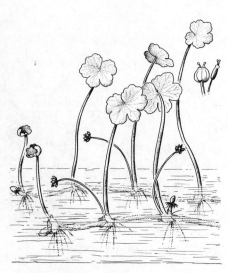

Hydrocotyle ranunculoides

are large with inflated stalks. The leaflets are sharply toothed and lobed. The
broad flat flower clusters (compound
umbel) are white, but the overall
aspect is greenish white against the
yellowish-green background of the
foliage.

MARSH PENNYWORT *(Hydrocotyle)*
Parsley Family

At margins of most any fresh pond or
lagoon that is not already crowded with
tules and other sedges, the water or
soggy mud will be completely concealed by a covering of the shining dark
green round leaves of the pennywort.
There are two kinds that grow on Point
Reyes Peninsula. The leaves of *Hydrocotyle verticillata triradiata* have
scarcely scalloped edges, and the stalks
are affixed to the leaf blade near its center instead of its margin, much like the
attachment of leaf to stalk in the garden nasturtium. The other, *H. ranunculoides,* as can be guessed by the
name, has a lobed leaf much like a buttercup *(Ranunculus),* and the leaf stalk
attaches as most plant leaves do to the
margin instead of the middle. The flowers and fruits are inconsequential and
will probably not be seen, as the stalks
that hold the small tight clusters of
greenish-white flowers are shorter than
the leaves. The species of this genus
are marsh and pond plants that grow in
both hemispheres.

WATER PARSLEY *(Oenanthe
sarmentosa)*
Parsley Family

This is a true aquatic that will grow
in the shallow water and mud of
ditches, as well as freshwater ponds or

lakes where it often branches and re-branches into great masses of leaves and flowers. The leaves are composed of toothed, sharp-pointed leaflets. The flowers are white; each flower is borne on stalks about an inch long, all arising from the top of a stem or branch. In botanical words, this is called an umbel. The stems are thick and juicy as is to be expected in a water plant, and roots will start up from the joints of the stem. In ditches or sluggish streams it often grows with watercress which also is white flowered, but in that plant the flowers are not in "umbels." Don't confuse the two. Some kinds of water-loving plants of the parsley family are poisonous.

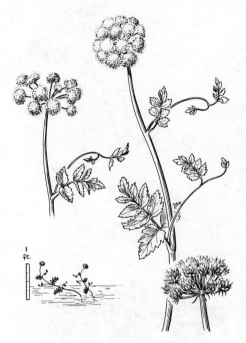

WATER PARSLEY

CALIFORNIA HONEYSUCKLE
(*Lonicera hispidula vacillans*)
Honeysuckle Family

California honeysuckle is a vine, and it can manage to reach a length of 20 feet in the shrubbery where it grows. One wonders how it can grow so far when it has no tendrils, nor does the stem really twine. The cordlike leafless stems hang loosely from the trees, and at the top are the hairy leaves, the flowers and eventually the fruits. The flowers are pink, without fragrance, and half the length of the honeysuckle flowers of the garden. The flowers are clustered in whorled spikes and usually sit bouquet-fashion against the pair of uppermost leaves which are fused into one. The berries, which ripen in fall, are a translucent bright red color.

CALIFORNIA HONEYSUCKLE

COAST TWINBERRY

BLUE ELDERBERRY

RED ELDERBERRY

COAST TWINBERRY (*Lonicera involucrata ledebourii*)
Honeysuckle Family

Along streamsides and boggy swales you will find the twinberry shrubs growing, probably together with red elderberry, salmonberry, and flowering currant. Like the red elderberry, these four- to five-foot shrubs lose their leaves in winter. The plants are more attractive in fruit than in flower, though the hummingbirds of course prefer the flowering stage. The tubular yellow flowers come in pairs, each pair set in cupped bracts at the tops of the flower stalks. The bracts, as the fruit ripens, turn a rosy red and set off the bug-eyed pair of shining black fruits to perfection.

BLUE ELDERBERRY (*Sambucus mexicana*)
RED ELDERBERRY (*Sambucus pubens arborescens*)
Honeysuckle Family

In the early spring the north-facing slopes are thickly dotted with shrubs of the red elderberry just coming into leaf and flower. The shrubs with their cream-colored lobed domes of blossoms set off by the light green expanding leaves can be identified at a good distance as easily as the blue-flowered mountain lilac. The ripened berries are bright red, and attractive but should not be eaten. When eaten raw they will certainly make you sick and probably would when they are cooked. If you feel the need of an elderberry pie or a glass of jelly you should make them with blue elderberries. These plants with their characteristic flat-topped sprays of flowers and fruit can be observed on the slope west of the head of Tomales Bay in late May and June, growing four to twelve feet high in the mixed shrubbery growth.

SEASIDE AGOSERIS or DANDELION
(Agoseris apargioides)
LARGE-FLOWERED AGOSERIS
(Agoseris grandiflora)
Sunflower Family

Seaside agoseris, which has coarse prostrate leaves, grows on the bluffs and dunes. The large yellow dandelionlike flower heads are very showy on a sunny day. The heads grow directly from the rootcrown on erect stems. Some of the leaves are cut nearly to the midrib, others are nearly entire or toothed near the base. There is variation, too, in the amount of hair on the leaves. Other varieties of the species have been described; all grow naturally on the immediate coast.

Large-flowered agoseris *(Agoseris grandiflora)* grows in grassland or brush. The stalks bearing the single flower heads are about ten to eighteen inches tall, and there is some cobwebby wool on the head and the partially erect leaves. The seeds are carried about in the wind by their "parachutes" and look very much like those of the dandelions of lawns and flower gardens.

LARGE-FLOWERED AGOSERIS

PEARLY EVERLASTING

BEACH SAGEBRUSH

PEARLY EVERLASTING *(Anaphalis margaritacea)*
Sunflower Family

Not all plants in the sunflower family look like sunflowers or daisies. Within the family limits, related genera are placed in groups called tribes such as the chicory tribe, the thistle tribe, and, as with *Anaphalis,* the everlasting tribe. The famous edelweiss is also of this group and to my mind a much less handsome plant. Stems of the pearly everlasting are a foot to a foot and a half tall, and in summer the flattish clusters of flower heads make their appearance. The bracts around the individual heads of flowers are gleaming white, not shining, and retain this quality when dry. The tops of the leaves are dark green; the undersides are white with felty hairs and so are the stems. The plant grows in favorable locations varying somewhat from place to place, in temperate regions around the northern hemisphere.

BEACH SAGEBRUSH or SAGEWORT *(Artemisia pycnocephala)*
Sunflower Family

Horticulturists have become aware of the decorative possibilities of this half-woody perennial of the beach and dune which is native from Monterey to Oregon. The small yellowish flower heads are arranged in spikes on the ends of the upturned branches, but the charm of the plant comes from its growth habit and its beautiful shining silver foliage which clothes the many-stemmed clumps clear down to the sand in which it grows.

Another kind of sagebrush *(Artemisia californica,* not illustrated), is com-

mon on the headlands in the shrubby
mixture of poison oak, sticky monkey-
flower, lupine, coyote bush, and California coffeeberry that are found on the
upper reaches of the headlands. This
rounded shrub, too, has gray foliage
which is made up of leaves with threadlike divisions, but they lack the beautiful silky sheen of the beach sagebrush.
It makes its presence known to the
hiker by the clean herby tang of the
bruised foliage.

COYOTE BUSH

CHAPARRAL BROOM (*Baccharis
pilularis consanguinea*)
COYOTE BUSH (*Baccharis
pilularis pilularis*)
Sunflower Family

This native, both the species and its
variety, is not one of the most beautiful shrubs, but it is certainly the commonest. The plant is occasionally also
called greasewood. It is as aggressive a
weed as such European invaders as the
yellow-flowered brooms (*Cytisus* and
Spartium) or gorse (*Ulex*)—that shrub
that has become such a menace—or
even wild oats which many think of as
a native. The plants are evergreen,
acquiring new leaves along with the
winter rains. The seed-bearing plants
and those bearing pollen are separate
individuals. The seeds with their dense
silky tufts ripen in late summer and
fall. The species (*Baccharis pilularis
pilularis*) makes dense, nearly prostrate leafy mats and often has smaller
leaves than the variety. It belongs on
old dunes and on grassy headlands
along the sea, but it tends to wander
elsewhere. It is so attractive a shrub
that it has been used as ground cover
in gardens. The variety (*B. pilularis
consanguinea*) which is taller, is every-

CHAPARRAL BROOM

ENGLISH DAISY

where within the Seashore—on open hillsides, ocean bluffs, or in roadside brush or dense forests. One was found with a trunk three inches in diameter. The knobby growth often seen on the branchlets is not a flower bud, but the gall formed by an insect, in this case, a fly.

ENGLISH DAISY *(Bellis perennis)*
Sunflower Family

Daisies are just as much at home in a rather neglected lawn as dandelions are, but their blooming season is considerably shorter. At Point Reyes National Seashore they are a welcome floral addition in early spring along damp wooded trails in Bear Valley and appear elsewhere in the Douglas-fir forests along with that other escapee from the flower gardens—the forget-me-not.

THISTLE *(Cirsium)*
Sunflower Family

All the flower heads of thistles are large and spiny and the hundreds of florets are laden with pollen so in the flowering season—from late May and June onward—the thistle plants are much visited by bees. Five of the ten species of *Cirsium* that have been reported from Marin County are known to grow at the Point Reyes National Seashore. The bumblebees and other bees have done much cross-pollinating in the area so that natural hybrids are of frequent occurrence.

THISTLE *(Cirsium occidentale)*
Sunflower Family

The cobweb thistle *(Cirsium occidentale)* is the most beautiful of the thistles at Point Reyes. The foliage is covered with a loose filmy whitish covering and the spiny bracts around the head are draped with white "cobwebs." The flowers of the head are a rich purplish red. The plants grow on the coastal bluffs or on the sandy soil behind the dunes. The branching plants are two to four feet high.

Cirsium occidentale

THISTLE *(Cirsium quercetorum)*
Sunflower Family

The brownie thistle is usually very low with many spiny green leaves, but sometimes grows to a height of two or two and half feet. The bracts around the dirty white flower heads are spine tipped but not strongly so. It will grow on open grassy slopes or openings in the forest or brush. Like *C. occidentale* the blooming period is at its best in midsummer.

BEACH-ASTER *(Corethrogyne californica obovata)*
Sunflower Family

Sometimes this plant is called the prostrate beach-aster to differentiate it from *Corethrogyne californica* but it is

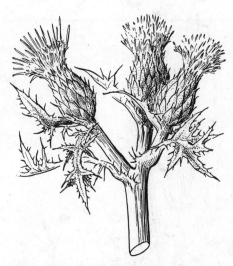

Cirsium quercetorum

BEACH-ASTER

not, strictly speaking, only a beach plant. In middle and late summer the single asterlike flower heads curve upward two to four inches from the tips of the creeping leafy stems. Plants are quite common on open grassy slopes in the bishop pine forest and the headlands above the ocean. The white woolly leaves are broad and rounded at the top and taper to a narrow neck where they join the stem. The lavender-rayed flower heads look very much like wild asters and differ only on a few technical points from those of the aster group that never grow tall.

BRASS BUTTONS (*Cotula coronopifolia*)
Sunflower Family

On the upper edges of the saltmarsh flats that circle parts of Drakes Bay and its estuaries there are broad patches of bright yellow brass buttons. However, the plant is not limited to the saline soil of tidal flats. It can be a semi-aquatic or a marginal plant at a freshwater lagoon or a sag pond. It is a plant that one gets to recognize in driving around the Bay Region, being particularly common in the flats above the saltmarshes by San Francisco Bay. Note the difference in sizes in the drawing. Plants can be an inch or so high in drying mud or several inches to a foot and a half tall in the tangle of vegetation at the shallow margins of the ponds. Few plants have been named so appropriately. The flower heads, packed tightly with many tiny blossoms, are the shape and size of brass buttons and show off very nicely against the succulent yellow-green foliage.

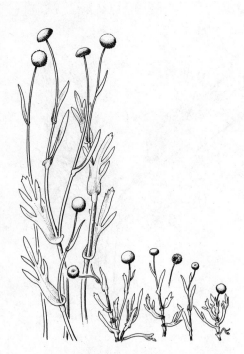

BRASS BUTTONS

SEASIDE DAISY *(Erigeron glaucus)*
Sunflower Family

The long blooming period of the seaside daisy during the spring and summer makes it one of the mainstays for color in the flower gardens of the ocean bluffs. The large flower heads, each one with about 100 petal-like rays around the yellow disk, are set rather closely against the leaves. The plants have a compact appearance. The leafy stems of about the same height, the clump of basal leaves, and the stout root give it the solidarity it must have to withstand the fierce coastal winds. Seaside daisy grows only on the California and Oregon coasts.

ERIOPHYLLUM or WOOLLY SUNFLOWER *(Eriophyllum lanatum arachnoideum)*
Sunflower Family

The eriophyllums are a western group of plants in which the different species are markedly dissimilar in appearance. In the genus one finds many types: annuals or perennials; those having small flower heads clustered like those of goldenrods, or plants with flower heads an inch across, including the yellow rays. Whether they are small or large the flowers are yellow. One species that might be found in the southern end of the National Seashore is called yellow yarrow *(Eriophyllum confertiflorum)*.

The stems and undersides of the leaves of the woolly sunflower *(E. lanatum arachnoideum)* are more or less woolly white with hairs and the upper side of the lobed or toothed leaves are gray green. The leafy stems

SEASIDE DAISY

WOOLLY SUNFLOWER

LIZARD TAIL

GUMPLANT

rise singly from the low perennial base and bear one or two attractive flower heads at the top. Look for the flowering plants in rocky banks or the poor soil on a hillslope in June and July.

SEASIDE WOOLLY SUNFLOWER or LIZARD TAIL (*Eriophyllum staechadifolium artemisiaefolium*) Sunflower Family

The variety of the lizard tail has leaves much like a sagebrush. The shrubby perennial is found along the coastal strand from Santa Barbara County to Oregon. Usually the shrubby plants are molded by the wind into one- or two-foot high mounds of gray foliage from which no branches spread out to spoil the symmetry. Even the stems bearing the short-stemmed clustered flower heads are quite short. It would be a fine evergreen shrub for a beach cottage. If the wind does not keep it pruned to garden size the pruning shears could.

GUMPLANT or GRINDELIA (*Grindelia stricta venulosa*) Sunflower Family

The coarse grindelia of the ocean bluffs and stable dunes is sometimes known as *Grindelia arenicola,* and is more frequently seen than *Grindelia maritima* of the coastal scrub and *G. humilis* of the saltmarshes, both of which are usually erect growing plants. Our gumplant is inclined to sprawl, and, though the branches may occasionally rise six to twelve inches from ground level, the plants should be measured in width instead of height. The toothed stem leaves are sessile on the stems and may be thin or thick. The yellow centered and rayed flower heads are quite large. When the whole head is still in the young bud stage, the center is white with a hard milky looking resin. The whole plant exudes a medicinally aromatic smell, and it has been used for medicine in comparatively recent times. A brew of the dried leaves and stems has been used to relieve bronchitis and asthma. By midsummer the flower heads make their best show.

SMOOTH CAT'S EAR

SMOOTH CAT'S EAR (*Hypochoeris glabra*)
HAIRY CAT'S EAR (*Hypochoeris radicata*)
Sunflower Family

The annual and the perennial cat's ears, even if they are merely weeds from Europe, lend a bright yellow color to the open grassy hills in late spring and part of summer. The hairy cat's ear (*H. radicata*) is found in the forest and bluffs by the beach. Near the beach it should not be confused with the seaside agoseris. Both have the typical dandelionlike flower, but the hairy cat's ear has four or five flower heads on a stalk instead of a single head. The floaters on the seeds are different, too. The smooth cat's ear is a smaller plant and seems to seed even more prolifically than its relative. It is this character of setting abundant viable seed that enables this plant to replace the less aggressive native wild flowers.

HAIRY CAT'S EAR

BEACH LASTHENIA

GOLDFIELDS

GOLDFIELDS *(Lasthenia chrysostoma)*
BEACH LASTHENIA *(Lasthenia macrantha)*
Sunflower Family

The large-flowered *Lasthenia macrantha* would seem to be, from its appearance, unrelated to the small-flowered, spring-blooming annual "goldfields" (*L. chrysostoma*) that paints large expanses bright gold on the thin-soiled grassy hills of central California. The coastal form of *L. chrysostoma* is occasionally found on the coastal bluffs of the Point Reyes Seashore. Beach Lasthenia is a short-lived perennial, but there is much variation to be found. One of the named varieties has very narrow leaves. All the plants in this group are variable and were formerly placed in the genus *Baeria,* but recent work has shown them to be insufficiently different from the genus *Lasthenia* with which they are now merged. Another species, *L. glabrata,* is quite common in alkaline soil of the saltmarshes.

BEACH LAYIA *(Layia carnosa)*
TIDY TIPS *(Layia platyglossa)*
Sunflower Family

In summer the aromatic perfume of sticky tarweeds growing in fields and rolling hills greets a Californian returning from a trip east, and even if he were blindfolded he would know that he was home. Not all tarweeds are summer bloomers. The appealing tidy tips which grow within the Point Reyes National Seashore on bluffs and grassy slopes comes as a spring annual and is a common wild flower elsewhere in California. Its inch-and-a-half wide flower heads with each petal-like yel-

BEACH LAYIA

low ray having a band of white is a great contrast to the beach layia, in which the white rays are so tiny that you have to stoop to see them. The beach layia has more numerous sticky glands than the tidy tips and consequently more odor to the foliage. It grows only on beaches and dunes from Northern California southward.

COAST MICROSERIS (*Microseris bigelovii*)
Sunflower Family

Coast microseris is a delicate-pale yellow-flowered annual that is apt to be abundant on the grass-covered hills only in the more rainy seasons. The blossoms are not large, perhaps 1/2-inch broad when all the little strap-shaped florets in the flower head expand in the sunshine. The plant belongs to the chicory tribe of the sunflower family. In the sunflower family what appears to be a single flower is a very specialized head of many flowers. The sunflowers, asters, or daisies have very tiny blossoms in a central disk and petal-like strap-shaped ones which are often of a contrasting color around the disk. In the chicory section of the family all the flowers of the head are strap-shaped like those of the familiar dandelion or the blue chicory. Coast microseris is not as abundant as the golden yellow smooth cat's ear which grows in the same kind of place later in the spring. In the local wildflower chronology they come after the shooting stars (*Dodecatheon hendersonii*) and with the pussy's ears (*Calochortus tolmiei*), and while you are on your knees admiring the fine detail in this charming member of the lily family you will probably see beside it the three- to six-

TIDY TIPS

COAST MICROSERIS

inch high microseris with its nodding buds and fruiting heads rising from the basal, often undivided, threadlike leaves. This species was another "first" in John Bigelow's collection in Marin County in 1854. It was first collected at Corte Madera and not on the Peninsula.

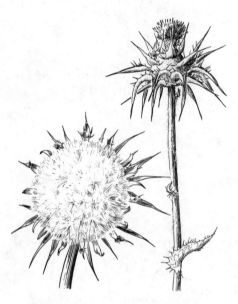

MILK THISTLE

MILK THISTLE *(Silybum marianum)*
Sunflower Family

If it were not an ever-present pest in pastures and along farmside roads, and if only a few of these three- or four-foot plants were to be seen, milk thistle might be considered a handsome plant. The large spiny leaves are marked with white. The large, reddish-purple flower heads are viciously spiny. With the efforts the dairy men make to wipe it out, it surely can't be forage for cattle. The pesticide applied does not kill immediately, and the plants put on a great effort of growth as shown by the elongated curving rubbery looking stems, and then the plants whiten and wilt and milk thistle is done for the year. There are other types of thistle *(Carduus pycnocephalus* and *C. tenuiflorus)* that are also local pests. Thistledown blows freely across fields in any one season as next year's crop plainly shows. The flower heads of the Italian thistle and the slender-flowered thistle, as these plants are called, are small and pink. Like most thistles they are spiny, but one would say they are irritatingly spiny rather than viciously so, as the spines are not as strong and stiff as those of milk thistle.

NARROW-LEAVED MULE'S EARS
(Wyethia angustifolia)
Sunflower Family

Regional history is often reflected in the scientific names given to plants. A trading expedition to the West was led by Nathaniel Wyeth in 1834. Wyeth collected and sent a kind of sunflower to his botanical friend Thomas Nuttall who had accompanied him on the journey west. Nuttall promptly found it to be new and described it as a new genus, Wyethia. At the Point Reyes National Seashore we have species different from the one that was first seen by Wyeth. Narrow-leaved mule's ears or compass plant, as it is sometimes called because the erect leaves (so it has been said) have their edges on a north-south line, is common on open grassy hillsides that have not been overgrazed. The plants grow in clumps, and in spring the broad (about one-and-a-half inches across) yellow flower heads show up plainly enough to be recognized as you drive by. The seeds of this and other sunflowers have been ground by Indians for food. Probably the local Indians used this species for that purpose.

NARROW-LEAVED MULE'S EARS

PUSSY'S EARS OR HAIRY STAR TULIP
(Calochortus tolmiei)
Lily Family

This is a modest relative of the peacock-eyed white, yellow, or brick-red mariposa lilies, which also bear the scientific name *Calochortus* and are found elsewhere in the West. So far this charming little *Calochortus* is the only kind found in the area of the Point Reyes National Seashore. By searching in grassy meadows lush with under-

PUSSY'S EARS

ground water, on a sunny spring day you will find these three-petaled, short-stemmed flowers. They have been called blue, but if they are it is a very muted blue. The petals are watery white, but the inside is thickly set with deep purplish hairs standing out essentially at right angles to the surface of the petal. They are textured just like the inside of a cat's ears, hence the common name. The three-angled seed pods are about one inch long.

MISSION BELL

MISSION BELL, CHECKER LILY, or RICE-ROOT *(Fritillaria lanceolata tristulis)*
Lily Family

Mission bells are widely distributed. In some form or other—for the plant is variable in size and color markings—it is found from Ventura County to British Columbia. The particular form occasionally found above the Point Reyes Beach or in the bishop pine forest could scarcely be called by the name "checker lily," as the surface of the "bells" are a bloom-covered brownish purple, through which the paler flecks of color so prominent in other varieties scarcely show. This is just another instance, as in the case of the yellow meadowfoam *(Limnanthes douglasii sulphurea)*, where the plant is just different enough to have a supplementary scientific name. In both of these cases the plants mentioned grow within the Point Reyes National Seashore and nowhere else. The name "rice-root" comes from the many fleshy scales resembling rice grains that are thrown off from the few thick cone-shaped scales that form the central core of the bulb. These bulblets grow only a single basal leaf until the bulb has attained sufficient size to grow a stalk of blossoms.

SLIM SOLOMON'S SEAL (*Smilacina stellata sessilifolia*)
Lily Family

In the Northwest this plant literally covers large areas in Washington and Oregon forests. It is not infrequent in Marin County and is common in the Point Reyes Seashore in the wind-molded mounds of shrubs and herbs that grow above the lagoons; equally common as a ground cover along Bear Creek and elsewhere. The dainty white flowers are not numerous and are arranged in a somewhat zig-zag fashion which more or less repeats the flat sprayed pattern made by the alternate leaves.

The fat Solomon's Seal (*Smilacina racemosa amplexicaulis*, not illustrated) is to be expected in the shaded slopes of the forests in the more southerly part of the Seashore. This variety is larger leaved and taller, and the many small white flowers are in a terminal dense-branched cluster.

SLIM SOLOMON'S SEAL

FALSE LILY-OF-THE-VALLEY
(*Maianthemum dilatatum*)
Lily Family

The two neatly ribbed heart-shaped leaves growing from the stem a little below the thick cluster of tiny white fragrant flowers make this plant look like a ready-made corsage. This northern-growing species makes its most southern appearance in San Mateo

FALSE LILY-OF-THE-VALLEY

CAMAS

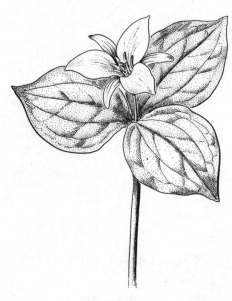

WAKE ROBIN

County. In the vicinity of Point Reyes it occurs commonly in the coastal grassland. Several pretty little red berries develop from the cluster of flowers, and they are reported to be edible. They are scarcely common enough, however, to offer much temptation.

CAMAS (*Camassia quamash linearis*)
Lily Family

The day is long past when the wet boggy meadows along the road to Point Reyes were "a sea of blue" in the spring with the flowers of the camas. It is to be especially deplored that the species is disappearing since the Point Reyes National Seashore is the southernmost point of its growth in the Coast Ranges. The spikes of short-stalked, star-shaped flowers may occasionally be found near the lagoons adjacent to the Point Reyes Beach. In the Pacific Northwest this and other kinds of camas were important food crops for the Indian tribes, who suffered a serious loss when the fertile land where the bulbs were abundant was plowed and planted with white man's crops.

WAKE ROBIN or WESTERN TRILLIUM
(*Trillium ovatum*)
Lily Family

This species of trillium blooms early in spring with a single-stalked white flower which rises as much as 1 1/2 inches above its three leaves. It is not common but can be found up Bear Creek in the shady woods on the slopes of Mt. Wittenberg and might be discovered on other north slopes in the Douglas-fir forests. Before the three white petals dry up they turn to pink or pale purple. The other trillium of Point Reyes Peninsula (*T. chloropetalum*, not illustrated), the mottled trillium or wake robin, has larger dark-spotted leaves. Its long, more slender petals may be green tinged with brown instead of the maroon or the greenish white colors that are typical in central and northern California. The flowers of the mottled trillium sit tightly against the three leaves. The plant is even less common than the western trillium, so if you see them growing report the locality to a ranger.

COAST ONION *(Allium dichlamydeum)*
Amaryllis Family

This coastal onion does not have a tall stalk, but the strong, rose-purple color of the flowers catches the eye as one walks over the headlands. It grows on open rocky slopes above the sea. Rumple the foliage and then sniff your hands, and you will know that this is as truly an onion as those you buy in the market. There are hundreds of kinds of onions growing in the northern hemisphere—about 50 different kinds in California alone—and besides the ones everyone knows, such as leeks, garlic, and shallots, there are several that make fine garden flowers if you don't crush the stems when you pick them for a bouquet. The sap always smells strongly of onions.

DWARF BRODIAEA *(Brodiaea coronaria macropoda)*
ITHURIEL'S SPEAR *(Brodiaea laxa)*
WILD HYACINTH, BLUE DICKS *(Brodiaea pulchella)*
Amaryllis Family

The bulblike corms of these members of the amaryllis family and their

COAST ONION

ITHURIEL'S SPEAR

DWARF BRODIAEA

WILD HYACINTH

DOUGLAS IRIS

many cousins around the state helped to keep various Indian tribes supplied with starchy food. Their disappearance from the landscape may be caused by too much cultivation by plows and bulldozers, though gophers will help ruin the corm crop if predators cannot keep the gophers under control. The violet-blue flowers of the wild hyacinth grow in tight clusters on their dark green, ten- to fifteen-inch stalks. Of the three that are pictured this one is the first to be found in bloom. *Brodiaea laxa,* called Ithuriel's Spear, is taller and has paler purple, thinner textured, funnel-like flowers. Another brodiaea *(B. peduncularis)* has fewer flowers and more white on the petals. The part of the flower that will become the seed-pod (ovary) is yellow instead of blue. This plant grows in marshy places. Another unillustrated species is *elegans,* the harvest brodiaea, which blooms in dry places in midsummer and resembles the following species in color and arrangement of flowers. The dwarf brodiaea *(B. coronaria* var. *macropoda)* is not so common, perhaps because it needs better drainage. Most of the stems are underground and the flower cluster of waxy purple flowers spreads from a stem that is scarcely an inch long. The leaves, as with the other species in the genus, are scarcely noticed, and they dry up very soon. It is often found in rocky soil of hilltops.

DOUGLAS IRIS *(Iris douglasiana)*
Iris Family

If the National Seashore were to have an official flower chosen from the multitude of kinds that grow there, none would be more appropriate than the

Douglas iris. In openings among the coastal scrub or in the windswept headlands where vegetation is only a few inches high, the clumps of deep purple iris are set about as effectively as if placed there by a landscape gardener. Neither deer nor cows seem to enjoy eating it. Away from the immediate coastline in the dark forest of Douglas-fir or among the bishop pines the stems are taller and not all the flowers are deep purple. Sometimes they are lavender with purple lines and have a white blotch at the base of the petals. They may even come down the color range to white marked with pencilings of purple. Clumps of them are especially effective in Bear Valley. They are at their best from late February to May, with a few occasionally blooming even later, since seasons on the Point Reyes Peninsula are not sharply marked. *Iris longipetala*, the coastal iris, which has no tube between the flower and seedpod-to-be and whose leaves are erect, may yet be discovered within the limits of the National Seashore. It grows on the other side of the San Andreas Fault.

BLUE-EYED GRASS

BLUE-EYED GRASS (*Sisyrinchium bellum*)
YELLOW-EYED GRASS (*Sisyrinchium californicum*)
Iris Family

The blue-eyed grass and the yellow-eyed grass are obviously close relatives, but one of the most marked differences between them other than color is the kind of place where each species is found growing. Yellow-eyed grass must have surface water,

YELLOW-EYED GRASS

REIN-ORCHIS

springs, swamps, or roadside seeps. The knife-edged dark green stems topped by two or three yellow flowers push their way to the light through masses of deer fern, sedges, and rushes on the edges of thickets of ledum, red elderberry, salmonberry, and flowering currant. Look for blue-eyed grass on open grassy hillslopes or the grassy headlands on the bluffs above the sea. Here the flat stems scarcely lift their deep purple-blue flowers two inches above the ground, for the wind sweeping in from the Pacific shears this and other herbs to tundra height. Away from the ocean the stems may be eight to twelve inches high.

REIN-ORCHIS (*Habenaria elegans maritima*) (*Habenaria dilatata leucostachys*)
Orchid Family

Calypso and lady slippers grow in Marin County, but are not yet known to occur in the National Seashore. More than one kind of rein-orchis does grow there and, though the flowers of these orchids are tiny, hundreds of them set thickly on the stem in a dense spike are much more impressive than a single medium-sized calypso, and they are fragrant besides. Look at a single blossom with a hand lens and you will find it as amazingly complicated as the orchids bought at florists.

The white rein-orchis (*Habenaria dilatata leucostachys,* not illustrated here), grows in wet swamps and has two to four green, pleated leaves on each stem. The only one pictured (*H. elegans maritima*), sometimes listed under the scientific name of *greenei,* has a very thick, two- to five-inch spike of white flowers, each spreading white segment with a green stripe. The stem leaves wither early. This grows on bluffs above the beach in drier places than does its relative, the white rein-orchis.

RUSH FAMILY
TO
GRASS FAMILY

The flowers that have been described so far in this book have been conspicuous mostly because of their color whether the flowers themselves were small, medium, or large. There is a vast number of plants lacking color, the majority consisting of the grasses, sedges, and rushes. These may tempt you to learn more about plants of the Seashore when you have mastered those with colored corollas.

First you should get a hand lens. Then go pick a grass with a fruiting or flowering stalk from the roadside, the barnyard or just pastureland. It can be seen immediately that it is texture and design that give beauty to the grass spikelet and individual grass floret. Take Kentucky bluegrass (*Poa pratensis*) for example and compare the accompanying illustration with the one you are looking at through your hand lens. Note the cobwebbed portion at the base of the bluegrass floret; the straight raised lines on the back (called nerves) with rows of tiny hairs on each one. Some other kinds of grasses have only one flower to a spikelet. That of panic grass has one fat seed with interesting hairs on one side and looks like something the canary would relish.

More than 150 kinds of grass grow in Marin County and a large percentage of these grow within the Point Reyes National Seashore. One of these, *Agrostis aristiglumis*, an annual, grows on gravelly slopes near Drakes Estero, and only in that spot. Grasses grow in all sorts of places: hilltops, meadows, ocean bluffs, freshwater or saltmarshes, in bishop pine forest or that of Douglas-fir or in the tanbark-oak, California bay, and madrone forests. Even the dunes have their population of grasses, some of which will not grow in any other habitat. For example, a species of wild rye is restricted to dunes. The handsome beachgrass called *Ammophila*, which was introduced from Europe as a sand binder, only grows in the dunes and is the commonest species there.

The sedge family is not so omnipresent as the grasses are in the Seashore. For the most part sedges are found in wet places whether they are saline or fresh and there are not as many kinds altogether as in the grass family. A few are part of the ground cover in forests.

There are even fewer rushes than sedges, and there is one kind, at least, that is conspicuous. Those tussocks of dark green on the open treeless slopes that one sees as one drives along, are plants of one of the rushes (varieties of *Juncus effusus*).

Whether it is a grass, a sedge, or a rush that you are studying with your hand lens, the accompanying sketches showing representatives of the three different plant families will help you. If you wish to go beyond that, consult John Thomas Howell's *Marin Flora* which lists all the plants that grow in the county and then if you wish to delve still further look up one of the California floras such as that written by Philip A. Munz and collaborators. This book covers the whole state and has descriptions of each plant species included. There are also books devoted to grasses only.

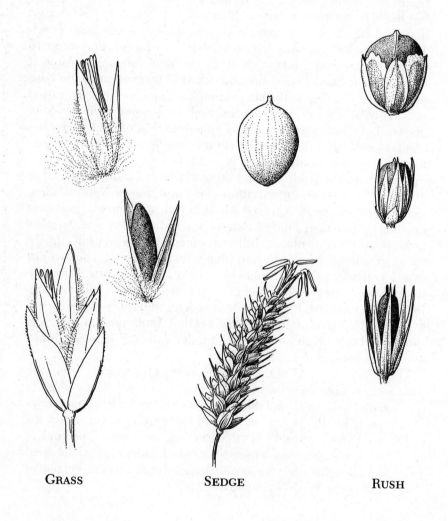

GRASS   SEDGE   RUSH

RUSH FAMILY

Only 15 species of this family are found in Marin County according to Howell's *Marin Flora,* and one of these is the woodrush (*Luzula multiflora*). The remaining 14 belong to the genus *Juncus,* the genus that gives the family its name.

BOLANDER RUSH (*Juncus bolanderi*)

The flowers of Bolander rush are clustered into a tight, round, brown head while those of the common rush are in an open arrangement. Bolander rush grows only in seeps and springs and is not uncommon in the Coast Ranges from central California north to British Columbia. This rush was named to honor Henry Bolander, the man who first collected it. He was appointed State Botanist in 1864 and held that post for several years. During his life-time he was a very active collector of plants from all over the state and bota-nists through the years have described as new quite a few species based on some of the collections he made.

COMMON RUSH (*Juncus effusus* varieties)

Common rush thrives best when its roots are damp. It will even grow in springs. Though it has not been proved, the local Indians may well have used the pliable but strong stems in their household chores. Their northern neighbors did and found them useful. Even today it has utilitarian uses. I have purchased, in markets, bunches of watercress tied together with the pliant stems of the common rush. Several varieties (or kinds) have been

BOLANDER RUSH

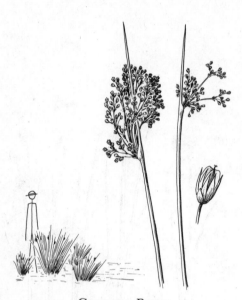

COMMON RUSH

LYNGBYE'S SEDGE

named and are commonly found growing here and other places in the United States but what botanists call the name-bearing variety (*Juncus effusus* var. *effusus*) is not believed to grow in North America.

SEDGE FAMILY

The differences between grasses and sedges become more apparent with a little practice. The diagrams at the beginning of this section and a few generalizations about these two plant families and the rush family will point up the differences between them. Grasses have hollow stems, solid only at the joints or nodes as they are called. Sedges do not have hollow stems, are not jointed and the stems are often three-sided. In grasses the flower parts, that is, the stamens, pistil, and ovary, are set between two bracts; in sedges the flower parts have but one bract (lacking in a few species) or sometimes a circle of bristles. Except that they are not colored the flowers of the rush family do not look structurally different from an onion or a brodiaea blossom and, instead of only one seed per flower developing as is true with grasses and sedges, the rushes have capsules for fruiting bodies in which many seeds develop.

On the Point Reyes Peninsula the genus *Carex* includes by far a larger number of species than do the other genera of the sedge family that occur here. Point Reyes Peninsula is the southern limit for a few of the species of sedge. The saltmarsh plant described below is an inhabitant of the Arctic seacoast and is able to exist this far south because of the favorable conditions for its survival at this latitude.

Lyngbye's Sedge (*Carex lyngbyei*)

This is a rare plant and you may not be lucky enough to see it growing. The saltmarshes and brackish places are its natural growing area and by searching in just the right habitat you may see the stems that bear the drooping blackish spikes rising out of the mud and just as much at home here as they would be in Kamchatka or Greenland. This plant differs from the common marsh sedge in that the stems which grow from the creeping rootstocks are rather evenly spaced instead of growing in clumps.

Slough Sedge

Slough or Common Marsh Sedge (*Carex obnupta*)

At least most of the springs and freshwater seeps in the Seashore will have large clumps of this common marsh sedge. The spikes are stalkless or nearly so (sessile) and are erect or spreading from the axis of the stiff erect stems. The stems overtop slightly the dense growth of leaves that form the clumps. The species is rather common along the coastal mountains from Alaska to San Luis Obispo County in California.

Bulrush or Common Tule (*Scirpus acutus*)

Although the kinds of *Scirpus* pictured together here, may not be the tallest and the shortest in the genus, they certainly are so on the Point Reyes Peninsula. The common tule is four to six feet high or even more; the low clubrush is four to six inches high or even less. The tule is a perennial with

Bulrush

LOW CLUBRUSH

Scirpus koilolepis

SILVER HAIRGRASS

few tall stems from a creeping underground stem and a cluster of inconspicuous short dry sheathing leaves. The
blooming part (inflorescence) of the
plant which occurs toward the summit
of the pithy stem is a cluster of brown
spikelets. There are several different
kinds of tules throughout the state and
elsewhere.

LOW CLUBRUSH *(Scirpus cernuus* var.
californicus)
(Scirpus koilolepis)

The low clubrush has many small
stems from the fibrous roots. Each stem
is topped by a small spikelet which
seemingly terminates the stem though
actually it is in the axis of a small leaf.
This species or forms of it are found in
wet places all around the world. A related low-growing species *(Scirpus
koilolepis)* also has a leaf at the base of
the spikelet which extends well beyond the flowering part and looks like
just *more* stem.

THE GRASS FAMILY

SILVER HAIRGRASS *(Aira
caryophyllea)*

This is one of the daintiest grasses I
know. It too, like the quaking grass, is
a European, but it is by now a thor-
oughly naturalized part of the landscape. The tiny spikelets (less than an
eighth of an inch long) each borne on
hairlike stalks, shine in the sun. As the
leaf blades are scarcely noticeable the
upper third or more of this annual plant
is devoted to flowering and with the
many shining spikelets is sometimes
used as a part of a "dry bouquet." But
remember! You are on national property and the plant should not be picked.

QUAKING GRASS (*Briza minor*)

The chances are very good indeed of
finding quaking grass and silver hairgrass growing with the small annual
festucas of open slopes and pathways.
You may even find along the pathways
and roadsides another species of Briza
(*B. maxima*) also a native of Europe,
called rattlesnake grass. This species
with large spikelets (see detail), is in-
vading the area. The spikelets on quaking grass are erect and the branches of
the inflorescence spread. The spikelets of rattlesnake grass are drooping
and the spikelets are much larger as the
drawing shows. Anyone can see that
they are close relatives, in other words,
related species of one genus.

QUAKING GRASS

NOOTKA REEDGRASS (*Calamagrostis nutkaensis*)

Probably the largest of the bunchgrasses on Point Reyes Peninsula is the
Nootka reedgrass. It is a species that
grows in moist places in coastal forests
and freshwater marshes from Monterey
to Alaska. It is abundant at Point Reyes
in ledum swamps and the drainage
gullies in the forests. The clumps reach
a height of three or five feet including
the flowering stems which bear hundreds of tiny florets in green plumes.
later turning dingy purplish. No grass
seeds are very large except some of the
cereals and Job's Tears but the seeds of
Nootka reedgrass seem unusually small
in relation to the huge plants they produce. Examine one of the seeds with a
very good hand lens. You will see a
short coarse hair arising from the back
of it (this is called an awn) and a tuft
of soft hairs at the base. If you are using
a key in a book which lists the different kinds of grasses you will learn from

NOOTKA REEDGRASS

2 ft.

CALIFORNIA FESCUE

VANILLA GRASS

these "discoveries" with your hand lens that you have identified a species of the genus *Calamagrostis*.

CALIFORNIA FESCUE (*Festuca californica*)

Another bunchgrass that is not as selective as vanilla grass about places where it can grow, is the California fescue. It is found, but not growing abundantly, most anywhere from Monterey County to Oregon (a variety of it grows in southern California) in moist swales, dry chaparral, yellow pine forest, or open woodland. The clumps are three feet or less in height. The stems and leaves are bluish-green and where the leaf blade and the sheath meet there is a collar of hairs that can be seen without a lens. The flowering branches as they first appear fit tightly against the main axis of the stem but soon spread widely almost at right angles. Several kinds of annual fescues are common on open grassy slopes and hilltops. These plants are six to twelve inches high, having single stems or with many stems arising from the base like tiny bunchgrasses. Only when you get out your hand lens and compare the flower heads will you see why the big perennial and the small annuals belong to the genus *Festuca*.

VANILLA GRASS (*Hierochloe occidentalis*)

Vanilla grass or California vanilla grass is one of the more handsome of the bunchgrasses in the forests of bishop pine or mixed forests of tanbark-oak and madrone or in stands of Douglas-fir. If there were redwoods in the National Seashore this grass would be growing in its spotty fashion under them too, because throughout the redwoods that range from Monterey north it is the most noticeable grass of those forests. The clumps are about two feet high and the stems are often red. The bright green leaves which are topped by the greenish tan and shining panicles of florets are attractive enough, but the added dividend is the odor of the leaves which when rubbed give off a delicate fragrance of, if not of vanilla, perhaps of new mown hay.

VELVET GRASS (*Holcus lanatus*)

In June when velvet grass begins to head out, the plants color fairly large patches a dusty purple (in the open woods or freshwater marshes) for velvet grass is well adapted to our foggy climate and grows profusely. They grow a foot and more high and are single stemmed. The panicle (flower-bearing part) is often purplish and the leaves and stems a soft sagelike green. The grass is pleasant to the touch as well as the sight. The leaf sheath around the stem, the leaf blades, and the stem itself are thickly set with soft spreading hairs. The plant feels like velvet when you grasp it.

CALIFORNIA BOTTLEBRUSH GRASS (*Hystrix californica*)

California bottlebrush is the most graceful of the large grasses in the area. As tall as some of the garden bamboos, six feet and less, its slender stems arise singly from joints on the underground running rootstock, thus differing from

VELVET GRASS

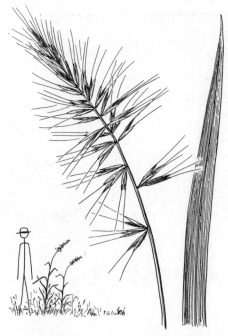

CALIFORNIA BOTTLEBRUSH GRASS

the reedgrass, California fescue, and vanilla grass which are bunchgrasses. It is found in shady woods, perhaps more frequently in the mixed thickets associated with the woods. The fruiting head, about as big as a head of barley, nods on the top of the stems above the shrubbery. In its young stages before it is in complete flower the "bottlebrush" effect is not so evident as it is later when the seeds ripen. Though not rare with us, California bottlebrush has a very limited range. It grows only in woods along the coast from Sonoma County to Santa Cruz County.

PACIFIC PANIC GRASS (*Panicum pacificum*)

One expects plants that flower from early spring to autumn to change their appearance somewhat through the months. California poppies that bloom in the spring and late summer on the same plant look very different in size and shades of color. It is a bit startling, however, to find a grass with a completely different growth form for the different seasons and growing in a climate, too, where there is as little seasonal change as there is on the Marin seacoast. In its natural habitat in moist meadows and streamlets the spring form of panic grass will have a basal tuft of leaves and the flowering stems with their tiny spikelets arise from the base of the plant. In summer and early autumn the stems lengthen and form a thick loose turf, becoming quite leafy and shorter-stemmed panicles grow from the leaf axils.

PACIFIC PANIC GRASS

RABBITFOOT GRASS (*Polypogon monspeliensis*)

No roadside ditch in the western states is complete without rabbitfoot grass. It is a European annual and like others, such as wild oats, quaking grass, and velvet grass, has made itself at home and become a useful addition for grazing. The drawing shows you the shape of the flowering head but it cannot show you its yellow-green color and shining furry appearance, silky soft to the touch.

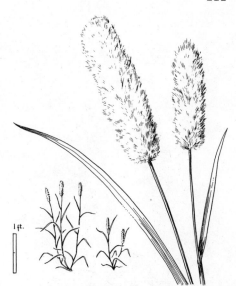

RABBITFOOT GRASS

PACIFIC CORDGRASS (*Spartina foliosa*)

To find this you must go to the salt-marshes bordering the estuaries. A freshwater lagoon will not do. Here along the network of channels washed by the tide and below the line of the pickleweed is your grass. The plants have strong creeping rootstocks; the stalks are stiffly erect and grow to three feet or less and the flowering stalk (inflorescence) is thick and is four or five inches long. Pacific cordgrass as the name suggests grows along the salt-marshes adjacent to the ocean from Canada to Baja California.

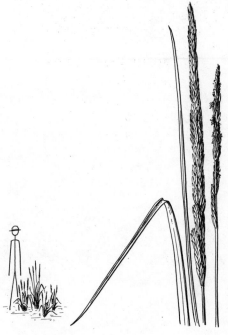

PACIFIC CORDGRASS

SELECTED REFERENCES

GRASSES

Chase, Agnes. *First Book of Grasses: The Structure of Grasses Explained for Beginners.* Washington, D.C.: Smithsonian Institution, 1959 (third edition).
Hitchcock, A. S., and Agnes Chase. *Manual of the Grasses of the United States.* Washington, D.C.: Government Printing Office, Misc. Pub. 200, U. S. Dept. Agr., 1951.
Pohl, Richard W. *How to Know the Grasses.* Dubuque, Iowa: William C. Brown, 1968 (second edition).

GENERAL

Abrams, Leroy and Roxana S. Ferris. *Illustrated Flora of the Pacific States.* 4 vols. Stanford: Stanford University Press, 1923-1960.
Jepson, Willis L. *Manual of the Flowering Plants of California.* Berkeley and Los Angeles: University of California Press, reprinted 1970.
Mason, Herbert L. *Flora of the Marshes of California.* Berkeley and Los Angeles: University of California Press, reprinted 1969.
Munz, Philip A. *A California Flora.* Berkeley and Los Angeles: University of California Press, 1959.
——. *California Spring Wildflowers.* Berkeley and Los Angeles: University of California Press, 1961.
——. *Shore Wildflowers of California, Oregon, and Washington.* Berkeley and Los Angeles: University of California Press, 1964.
——. *Supplement to a California Flora.* Berkeley and Los Angeles: University of California Press, 1968.
Parsons, Mary E. *Wildflowers of California.* New York: Dover, 1966.

LOCAL

Ferris, Roxana S. *Native Shrubs of the San Francisco Bay Region.* Berkeley and Los Angeles: University of California Press, 1968.
Howell, John Thomas. *Marin Flora.* Berkeley and Los Angeles: University of California Press, 1970 (second edition).
Metcalf, Woodbridge W. *Native Trees of the San Francisco Bay Region.* Berkeley and Los Angeles: University of California Press, 1959.

Sharsmith, Helen K. *Spring Wildflowers of the San Francisco Bay Region*. Berkeley and Los Angeles: University of California Press, 1965.

Smith, Gladys L. *Flowers and Ferns of Muir Woods*. Muir Woods Natural History Association, 1963.

Thomas, John H. *Flora of the Santa Cruz Mountains of California*. Stanford: Stanford University Press, 1961.

Young, Dorothy K. *Redwood Empire Wildflower Jewels*. Healdsburg, Calif.: Naturegraph, 1964.

INDEX